Philosophers in Depth

Series Editor
Constantine Sandis
Department of Philosophy
University of Hertfordshire
Hatfield, UK

Philosophers in Depth is a series of themed edited collections focusing on particular aspects of the thought of major figures from the history of philosophy. The volumes showcase a combination of newly commissioned and previously published work with the aim of deepening our understanding of the topics covered. Each book stands alone, but taken together the series will amount to a vast collection of critical essays covering the history of philosophy, exploring issues that are central to the ideas of individual philosophers. This project was launched with the financial support of the Institute for Historical and Cultural Research at Oxford Brookes University, for which we are very grateful.

More information about this series at
http://www.palgrave.com/gp/series/14552

Karim Dharamsi • Giuseppina D'Oro
Stephen Leach
Editors

Collingwood on Philosophical Methodology

palgrave
macmillan

Editors
Karim Dharamsi
Department of General Education
Mount Royal University
Calgary, AB, Canada

Giuseppina D'Oro
School of Politics, IR and Philosophy
Keele University
Staffordshire, UK

Stephen Leach
School of Politics, IR and Philosophy
Keele University
Staffordshire, UK

Philosophers in Depth
ISBN 978-3-030-02431-4 ISBN 978-3-030-02432-1 (eBook)
https://doi.org/10.1007/978-3-030-02432-1

Library of Congress Control Number: 2018960397

Cover illustration: © Artepics / Alamy Stock Photo
Cover Image: The Grey Tree (1912) by Piet Mondrian

This Palgrave Macmillan imprint is published by the registered company Springer Nature Switzerland AG
The registered company address is: Gewerbestrasse 11, 6330 Cham, Switzerland

In Memory of William Dray

Contents

Notes on Contributors

Jonas Ahlskog holds a PhD in philosophy from Åbo Akademi University (ÅAU, Finland) and defended his thesis on the philosophy of history in October 2017. Ahlskog is employed as a post-doctoral grant researcher in philosophy and history of ideas at ÅAU. His philosophical research is focused on questions concerning the human sciences, history, action, testimony, and cultural theory. In the history of ideas, Ahlskog is primarily doing research about conceptions of nation, class, language, and identity in political and ideological movements during the twentieth century in Finland. Ahlskog's historical research is part of an interdisciplinary research project about class, language, and national identity funded by the Kone Foundation. His articles have been published in journals such as *Rethinking History, Journal of the Philosophy of History*, *Clio* and *Historisk Tidskrift* (Sweden).

James Connelly is Professor of Political Theory at the University of Hull. He has authored numerous articles and books on the history of philosophy and environmental politics which have been published widely. His particular philosophical interests lie in the philosophy of the British idealists and, in particular, the philosophy of R.G. Collingwood. He has authored a monograph, *Method and Metaphysics: the Political Philosophy of R.G. Collingwood* (Imprint Academic, 2003) on the political philosophy of R. G. Collingwood; co-authored, with Peter Johnson and Stephen Leach, *R. G. Collingwood, A Research Companion* (Bloomsbury, 2015); and co-edited, with Giuseppina D'Oro, a revised edition of Collingwood's *Essay on Philosophical Method* (OUP, 2005). Many of his papers on the philosophy of Collingwood have been published widely.

Giuseppina D'Oro is Reader in Philosophy at Keele University. She is the author of *Collingwood and the Metaphysics of Experience* (Routledge, 2002). She co-edited, with James Connelly, Collingwood's *An Essay on Philosophical Method* (OUP, 2005), with Constantine Sandis, *Reasons and Causes: the Causalism/Anti-Causalism Debate in the Philosophy of Action* (Palgrave 2013), and with Søren Overgaard, *The Cambridge Companion to Philosophical Methodology* (CUP, 2017). She is the author of numerous journal articles on Collingwood's conception of the role and character of philosophical analysis and its continuing relevance to the argument for the autonomy of the human sciences. In 2015–2017 she was principal investigator (with Paul Giladi and Alexis Papazoglou) on a Templeton funded project: Idealism and the Philosophy of Mind.

Karim Dharamsi is Associate Professor of Philosophy and Chair of General Education at Mount Royal University in Calgary, Canada. He has written many articles on the philosophy of history and the philosophy of R.G. Collingwood which have been published widely. His articles 'Re-enactment in the Second Person' and 'From Norms to Uses and Back Again' have appeared in the *Journal of the Philosophy of History*. 'Mind as Action: Reflections on Collingwood and the Cause-Reason Divide' has appeared in *Anglo-American Idealism: Thinkers and Ideas* edited by James Connelly and Stamatoula Panagakou. He has published articles and co-edited collections on Wittgenstein, Frege, the philosophy of education, and liberal education.

Vasso Kindi is Professor of Philosophy at the Department of History and Philosophy of Science of the National and Kapodistrian University of Athens, Greece. Her articles on philosophy of science, Kuhn, Wittgenstein, Collingwood, philosophy of history, and conceptual change have been published widely. Her publications on Collingwood include the papers 'Collingwood, Wittgenstein, Strawson: Philosophy and description', *Collingwood and British Idealism Studies* (2016), and 'Collingwood's Opposition to Biography', *Journal of the Philosophy of History* (2012). She has authored, among others, the books *Kuhn and Wittgenstein: Philosophical Investigation of the Structure of Scientific Revolutions* (1995, in Greek) and *Kuhn's the Structure of Scientific Revolutions Revisited*, co-edited with Theodore Arabatzis (Routledge, 2012).

Chinatsu Kobayashi (Université du Québec à Montréal) holds a PhD in philosophy from the University of Ottawa, with a thesis on *Collingwood on Re-enactment: Understanding in History and Interpretation in Art*. She is writing another doctoral thesis in art history focusing on Ruskin's aesthetic theory and its influence on decorative arts. She is the author of a number of papers, including 'British Idealist Aesthetics, Collingwood, Wollheim and the Origins of Analytic Aesthetics' (*The Baltic International Yearbook of Cognition, Logic and Communication*, 2009), and, with M. Marion, 'Gadamer and Collingwood on Temporal Distance and Understanding' (*History and Theory*, 2011); and 'Heidegger, Japanese Aesthetics, and the Idea of a 'Dialogue' between East and West' (in W. Sweet ed., *Migrating Texts and Traditions*, 2012).

Stephen Leach is Honorary Senior Fellow at Keele University. He is the author of *The Foundations of History: Collingwood's Analysis of Historical Explanation* (Imprint Academic, 2009) and *A Russian Perspective on Theoretical Archaeology* (Left Coast Press, 2015), and, with James Connelly and Peter Johnson, *R.G. Collingwood: A Research Companion* (Bloomsbury, 2015). With James Tartaglia, he has edited *Richard Rorty: Mind Language and Metaphilosophy: Early Philosophical Papers* (CUP, 2014), *Consciousness and the Great Philosophers* (Routledge, 2017), and *The Meaning of Life and the Great Philosophers* (Routledge, 2018).

Mathieu Marion (Université du Québec à Montréal) holds a DPhil from the University of Oxford (1992), for a thesis later published as *Wittgenstein, Finitism and the Foundations of Mathematics* (1998). He has taught at the University of Ottawa from 1994 until 2003, and at the Université du Québec à Montréal since. He has published numerous papers on the philosophy of Wittgenstein, on the history of British and Analytic philosophy, including an early paper on 'Oxford Realism: Knowledge and Perception' (*British Journal for the History of Philosophy*, 2000) and 'Theory of Knowledge in Britain 1850–1950: A Non-Revolutionary Account' (*The Baltic International Yearbook of Cognition, Logic and Communication*, 2009). More recently, his studies on the history of logic, for example, with H. Rückert, 'Aristotle on Universal Quantification: A Study from the Perspective of Game Semantics' (*History and Philosophy of Logic*, 2016), have been published.

Elena Popa works on causality and its connection to action within the fields of philosophy of science (realism and objectivity), philosophy of psychology (causal reasoning, temporal reasoning), metaphysics (realism vs. projectivism), and history of philosophy. Her articles on causal projectivism and objectivity, and historical aspects of manipulability approaches to causation have been published in journals such as *International Studies in the Philosophy of Science, Collingwood and British Idealism Studies, Acta Philosophica Fennica* and *Australasian Philosophical Review*; further work on causality and psychological research is currently in progress. She holds a PhD from Central European University and is currently teaching philosophy at the Asian University for Women, a Liberal Arts university with a support foundation in Cambridge, Massachusetts, dedicated to the education of women from developing countries in Asia. She previously held short research fellowships in the Netherlands and the UK.

Jan van der Dussen is Professor Emeritus of Humanities at the Open University of the Netherlands. Besides three books and various articles in Dutch, his publications include *History as a Science. The Philosophy of R.G. Collingwood* (Martinus Nijhoff, 1981), revised edition (Springer, 2012), and *Studies on Collingwood, History and Civilization* (Springer, 2016). He is the editor of the revised edition of Collingwood's *The Idea of History* (OUP, 1993); and, with W.H. Dray, he is the editor of Collingwood's *The Principles of History and Other Writings in Philosophy of History* (OUP, 1999). He has also edited, with Lionel Rubinoff, *Objectivity, Method and Point of View. Essays in the Philosophy of History* (Brill, 1991).

Bernard Williams (1929–2003) was Knightsbridge Professor of Philosophy at Cambridge University, Monroe Deutsch Professor of Philosophy at the University of California, Berkeley, White's Professor of Moral Philosophy at Oxford University, Fellow of All Souls, Oxford University, and a Fellow of the British Academy. He is the author of *Morality: An Introduction to Ethics* (1972), *Problems of the Self* (1973), *Utilitarianism: For and Against* (with J.J.C. Smart, 1973), *Descartes: The Project of Pure Inquiry* (1978), *Moral Luck* (1981), *Utilitarianism and Beyond* (with Amartya Sen, 1982), *Ethics and the Limits of Philosophy* (1985), *Shame and Necessity* (1993), *Making Sense of Humanity* (1995), *The Great Philosophers: Plato* (1998), *Truth and Truthfulness: An Essay in Genealogy* (2002). There have also been several posthumously published volumes of Williams' essays, including *In the Beginning was the Deed* (2005), *The Sense of the Past* (2006), and *Philosophy as a Humanistic Discipline* (2006).

Abbreviations for the Works of R.G. Collingwood

A	*An Autobiography* (1939)
ARB	*The Archaeology of Roman Britain* (1930)
EM	*An Essay on Metaphysics* (1940)
EPM	*An Essay on Philosophical Method* (1933)
IH	*The Idea of History* (1946)
IN	*The Idea of Nature* (1945)
NL	*The New Leviathan* (1942)
PA	*The Principles of Art* (1938)
PE	*The Philosophy of Enchantment* (2005)
PH	*The Principles of History* (1999)
RP	*Religion and Philosophy* (1916)
SM	*Speculum Mentis* (1924)

CHAPTER 1

Introduction: The Armchair and the Pickaxe

Karim Dharamsi, Giuseppina D'Oro, and Stephen Leach

After a long period during which metaphilosophy was shunned as philosophers chose to focus instead on first-order philosophical problems, reflections on the method of philosophy are once again occupying centre stage (Williamson 2007; Chalmers et al. 2009; Overgaard et al. 2013; D'Oro and Overgaard 2017). Collingwood was the author of two explicitly metaphilosophical treatises, *An Essay on Philosophical Method* (1933) and *An Essay on Metaphysics* (1940), and regarded the question "what is philosophy?" to be part of philosophy: "Reflection on it is part of itself" (EPM: 1). There has therefore never been a better time to revisit Collingwood's conception of the role and character of philosophical analysis and assess where he stands on a spectrum of views that range, as in the title of a recent collection, from the armchair to the laboratory (Haug 2013). Clearly Collingwood rejected both these extremes: he was neither the kind of metaphysician who conceived of philosophy as a form

K. Dharamsi (✉)
Mount Royal University, Calgary, AB, Canada
e-mail: kdharamsi@mtroyal.ca

G. D'Oro • S. Leach
School of Politics, IR and Philosophy, Keele University, Staffordshire, UK
e-mail: g.d'oro@keele.ac.uk; s.d.leach@keele.ac.uk

K. Dharamsi et al. (eds.), *Collingwood on Philosophical Methodology*, Philosophers in Depth,
https://doi.org/10.1007/978-3-030-02432-1_1

of armchair science seeking after ontological truths through reflection alone, nor did he think that philosophers could abandon the reflective method in favour of the experimental one.

But where exactly did he stand? Would he have been sympathetic to the Quinean view that philosophy is continuous with natural science? *Or* would he have taken the more moderate Lockean view, recently revived by Jackson (1998), that philosophy is an underlabourer to science whose task is to establish what aspects of the manifest image are compatible with the scientific image? While the editors of this collection are united in thinking that Collingwood had a much more robust sense of the role of philosophical analysis (and it may be better to declare one's views from the outset, rather than smuggle them in as a fact of the matter), the literature is quite divided on this issue and it is not always easy for the reader to navigate their way round it. Some think that Collingwood was indeed in the business of liquidating philosophy, even if the preferred special science in which he thought it should be dissolved was not physics but history. On this view Collingwood conceived of philosophy not as a normative but as a purely descriptive undertaking, whose task is to describe the belief systems of different people at different times and places. Metaphysics, on this view, is superseded not by physics but by cultural anthropology. This is still an influential view of Collingwood's philosophy with a long pedigree (Donagan 1962; Rotenstreich 1972; Toulmin 1972), and one that is shared by Williams (2006) (reprinted in this volume). Others have thought that Collingwood had a more robust sense of the distinction between philosophy and the special sciences and that he saw philosophy as distinct from them: rather than being continuous with any of the special sciences philosophy is in charge of excavating the presuppositions on which they rest. On this view there are two levels of investigation. The first-order level is that of the special sciences, which are the laboratories of knowledge; the second-order level is that of metaphysics, whose task is to make explicit the presuppositions that are entailed by the questions that are characteristically asked (and answered) in the sciences. While the metaphysician does not dictate to the historian or the physicist what their method should be, metaphysical analysis does make explicit to them what their method is, what is their distinctive *explanandum*, and how the subject matter of a special science differs from that of other forms of inquiry. The task of excavating these presuppositions does not therefore consist in describing what people believe (as if philosophy were a form of cultural anthropology) but what presuppositions are entailed by the characteristic questions asked in a given form of inquiry. In other words, on this reading, the task of metaphysical analysis is to uncover the presuppositions that are constitutive of a given form of knowledge and are mandatory for its practitioners. The consequence of this reading is that, far from being science's underlabourer (the Lockean view) or being continuous with science (the Quinean view), metaphysics is an autonomous form of inquiry whose subject matter are the presuppositions which are constitutive of the knowledge claims arrived at in the special sciences. To be clear, this reading of Collingwood's metaphysics of absolute presuppositions does not mean that he was in the business of defending the Cartesian view of metaphysics as an ontologically first science, lying at the roots of the tree of knowledge whose trunk is physics, and from whose branches (the special sciences) hang the fruits of knowledge which humanity enjoys (health and technological advances). Metaphysics, on our reading, is a first science only in the order of logical priority because it uncovers those presuppositions which the practitioners of the special sciences must make, as a matter of logical necessity, in order to engage in the production of knowledge. To read Collingwood in this way is to acknowledge a distinction between the modus operandi of the special sciences and that of philosophy: the former are involved in the production of knowledge; the latter enables one to understand the conditions on which knowledge rests, rather than produce a different kind of (metaphysical) knowledge.

The divide between these interpretative lines is often traceable to whether or not one thinks that *An Essay on Metaphysics*, which is normally read as advocating the dissolution of philosophy into history, is continuous with the earlier *An Essay on Philosophical Method* or whether it marks a historicist departure from it. A discussion of this issue is beyond the scope of this introduction but for the discontinuity thesis, which often goes hand in hand with the claim that Collingwood sought to liquidate metaphysics into history, the reader might consider Donagan (1962), Rotenstreich (1972), and Toulmin (1972). For the continuity thesis, which tends to go hand in hand with the claim that Collingwood did not relinquish a commitment to philosophy as an autonomous discipline with its own method and domain of inquiry, the reader might look at Connelly (1990) and D'Oro (2002). What is at stake between those who see discontinuity between *An Essay on Philosophical Method* and *An Essay on Metaphysics* and those who do not is the status of the a priori in Collingwood's later philosophy, and whether the notion of the a priori that is captured by the claim that knowledge rests on presuppositions of a certain kind (absolute presuppositions) is sufficiently robust to uphold the autonomy of philosophy or whether on the other hand it is weak enough to support the claim that philosophy is either a form of cultural anthropology or perhaps continuous with it.

An Essay on Metaphysics defends the view that all knowledge rests on some presupposition or other and that certain presuppositions, which Collingwood calls "absolute", are constitutive of a particular kind of knowledge in the sense that the form of knowledge which they support would not be possible without them. The principle of the uniformity of nature, for example, is implicit in any attempt to extrapolate general laws from particular observations and is thus presupposed by any attempt to advance empirical knowledge. The principle of the uniformity of nature is therefore absolutely presupposed by the making of inductive generalizations and constitutive of the kind of knowledge that natural science yields.

The view that the presuppositions on which knowledge rests are constitutive of the form of knowledge which they enable already marks an important departure from a strong (Kantian) notion of the a priori, where "a priori" means valid at all times and places. But how weak or robust is Collingwood's constitutive notion of the a priori? On the one hand one could argue that since metaphysics, as Collingwood conceives it, begins from knowledge as "we" have it, and since the "we" changes with the location of the subject in space and time, the presuppositions that metaphysics uncovers are historically parochial rather than universal and valid at all times and places. On the other hand, one could argue that although the notion of the a priori captured by the view that absolute presuppositions are constitutive of forms of inquiry is weaker than the Kantian notion of the a priori, it is not reducible to a mere (time-relative) historical a priori since it makes a claim that is stronger than "this is what 'we' in a given spatio-temporal slice, believe to be the case": the principle of the uniformity of nature, for example, is not merely something that we (here and now) believe to be true. It is rather a principle that anyone engaged in empirical enquiry is committed to, whether they are aware of it or not. Since the principle of the uniformity of nature is entailed by the advancing of inductive generalizations, anyone, at any time, who extrapolates an empirical law from particular observations can do so only by presupposing the principle. Be that as it may, as Williams (in the chapter reprinted here) says, there is often an ambiguity in the way in which philosophers who abandon foundationalism use the word "we" and it is this ambiguity that interpreters trying to make sense of the notion of the a priori at work in *An Essay on Metaphysics* must grapple with and try to resolve.

This interpretative divide has substantive consequences for how one understands the sense in which Collingwood belongs to the historicist tradition. For the term "historicist" has been used in very different ways. It has often been used as a term of abuse to indicate the willingness of a philosopher to espouse a form of epistemic relativism according to which what is true (or false) is relative to what a person or group of people believe to be true (or false). The view that Collingwood is a historicist in this sense (in the sense of being an epistemic relativist) is distinct from, and much stronger than, the claim that Collingwood is a historicist in so far as he held the view that in order to understand people who do not share our own assumptions we need to understand what they say and their actions on their own terms rather than our own. Stating that Collingwood was a philosopher sensitive to the historical context of action (something for which he is often praised) does not entail ascribing to him a commitment to the stronger claim that "x is true" means, or is synonymous with, "S believes x to be true". In the first sense, the attribute "historicist" is often used to describe Collingwood's philosophy in a pejorative way, to indicate a commitment to epistemic relativism, the view that the predicate "is true" means "S believes x to be true". In the second sense, the attribute "historicist" has been used mostly as a term of praise, to mean something quite different, namely that Collingwood warned against historical anachronism.

Collingwood was clearly a historicist in the second sense. He claimed that understanding past agents requires understanding the thought context of their action and was adamant that the historian should be sensitive to the differences between her own thought context and that of the agent. Historical inquiry, and the possibility of historical understanding, he claimed, rests on presuppositions that are very different from those informing the investigation of nature. The empirical investigation of nature, as we have seen, rests on the presupposition of its uniformity; the historical investigation of the past, on the other hand, rests on a very different presupposition, namely that the belief systems of historical agents may not be constant. Investigating the historical past as the natural scientist investigates nature (by assuming its uniformity) leads straight to committing the kind of historical anachronism Collingwood warns against. But the claim that in order to avoid anachronism in the study of past cultures the historian must make presuppositions which are very different from those at work in the natural sciences does not entail the stronger claim that all sciences are historical, that is, that they share the same presuppositions as the study of history. Natural science, as we have seen, is not a historical science precisely because the key presupposition natural scientists make when investigating the natural world is that it does not change.[1] Historicism, in the first and stronger sense, requires that the presupposition of natural science, the principle of the uniformity of nature, is a time-relative assumption rather than a logical condition that is entailed by the making of empirical generalizations (whenever or whoever makes them). And so it appears we have come full circle back to the question that has dogged Collingwood's scholarship: what is the status of absolute presuppositions? Are they logical conditions? Or are they historical ones? In what sense, if any, can they be both? While it would be overly optimistic to hope that the contributions to this edited collection will settle that question once and for all, they will hopefully enable the reader to understand what is at stake in answering that question one way or the other.[2]

1 Bernard Williams: An Essay on Collingwood: [Republished from "The Sense of the Past" (2006)]

There is general agreement that Collingwood's mature philosophy can be dated from *An Essay on Philosophical Method* (1933). In his *Autobiography* (1939) he deemed this to be "my best book in matter; in style, I may call it my only book" (A: 118). Most of those who are interested in Collingwood's philosophy share Collingwood's own high opinion of *An Essay on Philosophical Method*. Bernard Williams is exceptional in that although he has much admiration for Collingwood's work, in his posthumously published "Essay on Collingwood" he dismisses *An Essay on Philosophical Method* as "in fact mostly a dull and dated book, full of what are likely to seem unhelpful distinctions and assimilations". What really attracted Williams in Collingwood's work is Collingwood's writings in the philosophy of history and his discussion of the relationship between philosophy and history. One perennial misunderstanding that has dogged the reception of Collingwood's philosophy of history is the idea that Collingwood wishes to provide a prescriptive methodology for historians. Williams is fully aware that this is not the case and he argues against this misunderstanding both in this article and in his article on Collingwood in the *Dictionary of National Biography*, co-authored with Stefan Collini (Collini and Williams 2004). It is the discussion of absolute presuppositions in *An Essay on Metaphysics* that captivates Williams. Collingwood, Williams argues, tended towards a "radical historicism"; like many other philosophers, including Wittgenstein, who eschewed foundationalism, Collingwood used the word "we" in an evasive way, both in an inclusive sense "as implying universalistic preconditions on interpretation and intelligibility" and in a contrastive sense "under which 'we' here and now are distinct from others elsewhere and elsewhen, who lived in others and different intelligible human formations". The nuances of the relationship between philosophy and history, and the tensions between the local and the universal "we" are explored by many of the writers in this volume. Since Williams penned this essay much has been written on the relation between *An Essay on Philosophical Method* and *An Essay on Metaphysics*. As Collingwood's conversion to radical historicism in *An Essay on Metaphysics* has been put under some pressure, Collingwood's account of the nature of absolute presuppositions and of the role of conceptual analysis in metaphysics has been shown to be no straightforward form of historical relativism. It is to Williams' credit to have recognized Collingwood as a great, worthy of inclusion as a major figure in the history of western philosophy, his philosophical reservations notwithstanding.

2 James Connelly: The Development of Collingwood's Metaphilosophical Views

This chapter is an introduction to Collingwood's metaphilosophical views concentrating on the development of Collingwood's conception of philosophical methodology. It tracks his statements on metaphilosophy and method and goes on to ask whether it is appropriate to speak of an early and late Collingwood. In so doing it examines Collingwood's engagement with Oxford realism, logical positivism and his attempt to rehabilitate metaphysics. It locates Collingwood at the crossroads between the declining tradition of British idealism and the rising school of analytical philosophy.

It asks what was the nature of Collingwood's reconceptualization of the metaphysical task? Is there a danger that it leads to a dissolution of philosophy into history or is his later work part of a sustained effort to vindicate the autonomy of philosophical enquiry as distinct from the natural sciences? Its goal is to provide readers who are not acquainted with Collingwood's work with an account of the evolution of his metaphilosophical views thereby enabling them to navigate their way round the various disputes which surround it.

3 Jan van der Dussen: Collingwood on the Relationship Between Metaphysics and History

In this chapter Van der Dussen reads *An Essay on Metaphysics* in the light of an essay Collingwood wrote ten months earlier, “Function of Metaphysics in Civilization”. Van der Dussen claims that in *An Essay on Metaphysics* Collingwood does not so much argue for as assume that metaphysics is a historical discipline. Van der Dussen has sympathy for this position. He argues that the metaphysician uncovers presuppositions, but it is the task of the special sciences to change them.

He argues that it is only in relation to other metaphysicians that the Collingwoodian metaphysician can assume a critical role as opposed to a purely descriptive role. When not engaged in disputes with other metaphysicians, the Collingwoodian metaphysician works essentially as a historian.

By way of his own disputes with other metaphysicians (Krausz, Beaney, and Martin) Van der Dussen puts forward the thesis that Collingwood’s central argument in *An Essay on Metaphysics* is that metaphysics is a historical science. However, it is only the first four chapters of *An Essay on Metaphysics*, in which Collingwood develops his theory of presuppositions, that is philosophical. The rest of the book in which he makes use of this theory is historical. As such the book itself exemplifies what Collingwood advocated in his *Autobiography*, the rapprochement of philosophy and history.

4 Vasso Kindi: Presuppositions and the Logic of Question and Answer

Kindi explains that in Collingwood’s view the logician should study not just propositions but questions and answers. Logic, of the sort favoured by Russell and Whitehead, makes the mistake of studying propositions in isolation, whereas they should be studied in relation to questions. It is only when two propositions can be seen to answer the same questions that they can be said to contradict one another. This is apparent when we turn our attention to the special sciences. Collingwood tells us that he was led to these conclusions by his experience of archaeology.

> Meaning, agreement and contradiction, truth and falsehood, none of these belonged to propositions in their own right, propositions by themselves; they belonged only to propositions as the answers to questions: each proposition answering a question strictly correlative to itself. (A: 33)

We grasp the meaning of a proposition when we understand the meaning of the question which it attempts to answer. In that way we come to know more clearly that which we already know, but dimly. The metaphysician's task is to "excavate" different complexes of questions and answers until the absolute presuppositions which govern them are reached. It is this methodology that gives Collingwood's metaphysics its distinctive character. It is not reduced or assimilated to natural science or any other discipline but nor does it lose sight of the particular and the concrete. Kindi points out that here there is a similarity to the later Wittgenstein.

5 Elena Popa: Collingwood, Pragmatism, and Philosophy of Science

Popa investigates the relationship between Collingwood's metaphysics of absolute presuppositions and pragmatism. She notes that pragmatism is a broad church and that while there are family resemblances between Collingwood's project and that of philosophers of a pragmatist orientation, there are also some important differences. Collingwood tends to have in common with all pragmatists a rejection of the correspondence theory of truth, but unlike the classical pragmatists, he does not develop a positive theory of truth: his metaphilosophical strategy is to reject the view that absolute presuppositions are truth-evaluable rather than articulate an alternative conception of truth. Collingwood has therefore less in common with the classical pragmatists and (subject to a number of important qualifications) he is closer to later incarnations of pragmatism, in particular Putnam's internal realism and Price's discussion of causation as a form of explanation that is context sensitive. Since Collingwood's rejection of traditional metaphysical problems tends, like Price's, to be motivated by a pragmatics of explanation, Collingwood's rejection of the correspondence theory of truth does not lead to a full-blown deflationary strategy: rather than seeing philosophical problems as fake problems and advocating the end of philosophy, he sees the role of philosophical analysis to be that of making explicit the absolute presuppositions which inform different causal explanations in different explanatory contexts.

6 Giuseppina D'Oro: Why Epistemic Pluralism Does Not Entail Relativism—Collingwood's Hinge Epistemology

In her contribution D'Oro asks whether Collingwood's metaphysics of absolute presuppositions leads to the kind of belief-system relativism which is the target of Boghossian's sustained criticism in his *Fear of Knowledge* (2006). D'Oro argues that Collingwood's metaphysics of absolute presuppositions aims to defend a form of epistemic pluralism which is not reducible to the kind of belief-system relativism Boghossian critiques. Epistemic relativism claims that "x is true" means "S believes x to be true". But presuppositions for Collingwood do not do their logical work in virtue of being believed to be true (or false). Rather, they have "logical efficacy" in virtue of being entailed by the questions to which they give rise. Since Collingwood does not defend a form of epistemic relativism (not even belief-system relativism) he is not vulnerable to the self-undermining objection that is often raised against it. The decoupling of epistemic pluralism from epistemic relativism rests on a reading of absolute presuppositions as epistemic "hinges" which give rise to the characteristic complexes of questions and answers operative in different contexts of inquiry. The task of the metaphysician is to show that the questions asked in different contexts of inquiry are entailed by different absolute presuppositions that are constitutive of those forms of knowing. Epistemic pluralism rests on a form of hinge epistemology which denies that the predicate "is true" applies to the conditions of knowledge. But since epistemic pluralism is not a form of relativism, it is not vulnerable to the stock objections raised against relativism.

7 Karim Dharamsi: Oscillation and Emancipation—Collingwood on History and Human Nature

Dharamsi considers Collingwood's defence of the autonomy of the mental and contrasts it with the one articulated by liberal naturalists such as McDowell. Both Collingwood and McDowell, Dharamsi argues, acknowledge the irreducibly normative (in Collingwood's words: criteriological) nature of the study of mind and both reject the widespread naturalist assumption that philosophy is continuous with natural science. The liberal naturalist's and Collingwood's strategy are however fundamentally different. McDowell's liberal naturalism seeks to ease the tension between nature and mind through an Aristotelian account of how human beings acquire a second (rational) nature via the process of habituation into a culture. McDowell's strategy is therefore to soften naturalism so as to accommodate within its womb the normative character of the mental which a harder or more traditional form of naturalism struggles to provide a home for. From McDowell's perspective the problem of locating the mind in nature is a problem for traditional, harder forms of naturalism. Once naturalism is understood in a different way the problem that traditional philosophy of mind struggles with is eased. Collingwood agrees with McDowell's diagnosis of the problem, but not with his proposed solution. The problem of how the normative or criteriological aspect of the mind can fit in nature is indeed a problem for the hard naturalist, and solving the problem does indeed require, as McDowell advocates, moving beyond the underlying metaphilosophical picture of philosophy as continuous with the natural sciences. But the solution, for Collingwood, lies not in liberalizing nature, but in rejecting a conception of metaphysics as a science of pure being and understanding it instead as a historical enquiry into the presuppositions of science, including natural science.

8 Jonas Ahlskog: Collingwood and the Philosophy of History—The Metaphilosophical Dimension

Collingwood believed that our understanding of the philosophy of history matters for our understanding of philosophy as a whole. He did not completely spell out his philosophy, but it is possible to extend the trajectory of his thought.

In his unfinished *Principles of History* (1999) Collingwood is concerned with elucidating the a priori concepts (absolute presuppositions) that guide history, but he is also concerned with elucidating the relation between historical and philosophical thinking.

History, for Collingwood, designates a distinctive way of understanding reality, which involves "re-enacting" the motivational premise underlying an action. It is the philosophy of history that makes clear the distinctive character of history; and the philosophy of history is itself part of Collingwood's "metaphysics without ontology" which attempts to discover and make explicit the underlying presuppositions of different disciplines.

Collingwood's understanding of history as an autonomous discipline is contrasted to Hempel's view, but its interest is also pointed out in relation to more recent work in the philosophy of history that focuses on narrative and retrospective description.

In Collingwood's concern to distinguish the human sciences from the natural sciences there is common ground with Dilthey, Gadamer, and von Wright. However, Collingwood also worked on the relationship between history and philosophy. He wished to liquidate a particular conception of history, one that searches for timeless essences or confines itself to the analysis of pure logical form. This conception ignores philosophy's purpose of providing us with self-understanding.

History also gives us self-understanding, but our knowledge of how we acquire that knowledge is given to us by philosophy (itself in turn nurtured by an awareness of history).

9 Chinatsu Kobayashi and Mathieu Marion: The Later Collingwood On Method: Re-Enactment and Abduction

In this chapter Kobayashi and Marion first provide reasons to reject the many readings of Collingwood that sought to draft him as a participant in the Hempel-Dray debate about the status of covering laws in history. After all, this debate was not part of Collingwood's context and, although one can pry from his writings a contribution to it, one may simply, by doing so, misunderstand what he was up to. In the second part, they present the Gabbay-Woods Schema for abductive reasoning, as it occurs in the context of inquiry, as triggered by an ignorance problem, and as being "ignorance preserving". They then argue that this allows us better to see the point of Collingwood's "logic of questions and answers", as derived from his own practice in archaeology, and his use of the "detective model of the historian", as opposed to merely focussing on understanding what "re-enactment" could mean as a contribution to the Hempel-Dray debate.

10 Stephen Leach: Collingwood and Archaeological Theory

Leach asks what would Collingwood have thought of archaeological theory, a sub-discipline of archaeology that has developed since the 1960s and 1970s. He argues that Collingwood would have welcomed it for it has developed out of respect for the principle that in any investigation, in examining the evidence, in order for it to yield any answer one must always have some question in mind. Archaeological theory discusses what questions are worth asking.

Furthermore, Collingwood would have urged metaphysicians to take notice of such developments for not only did he leave his analysis of the presuppositions of history and archaeology unfinished but also, as he acknowledged, disciplines change and develop. (One example of an area in which there is still work to be done is the exploration of possible differences between history and archaeology. These differences might be highlighted by the use of Collingwood's theory of presuppositions.)

Nonetheless, although Collingwood would have welcomed the development of archaeological theory, and although he would urge philosophers to follow his example and to stay abreast of developments outside of their own discipline, he is still himself primarily a metaphysician rather than an archaeological theorist.

As such, in pursuing his own distinctive form of metaphysics, Collingwood does not shun the philosopher's study, nor the philosopher's armchair, but in his case the armchair is always placed next to the window.

Notes

1. *But please note that denying that natural science is a historical science is not the same as asserting that natural science is presuppositionless!* That is a completely different claim, one that Collingwood would have adamantly rejected on the grounds that there is no such thing as presuppositionless knowledge. Every form of knowledge has its own *explanandum*, which is the correlative of its method and presuppositions. The *explanandum* of natural science is events, which are known through the experimental method. The *explanandum* of history is actions, which are known through the historical method. There is no *explanandum* and no knowledge from nowhere.
2. Those interested in Collingwood's philosophical methodology might also be interested in this volume's elder sister "Collingwood and Philosophical Methodology", a special issue of *Collingwood and British Idealism Studies* edited by Giuseppina D'Oro and James Connelly.

Bibliography

Chalmers, D., D. Manley, and R. Wasserman, eds. 2009. *Metametaphysics*. Oxford/New York: Oxford University Press.

Collini, S., and B. Williams. 2004. R.G. Collingwood. In *The Dictionary of National Biography*. Oxford: Oxford University Press.

Connelly, J. 1990. Metaphysics and Method: A Necessary Unity in the Philosophy of R. G. Collingwood. *Storia, Antropologia e Scienze del Linguaggio* 5 (1–2): 36–156.

D'Oro, G. 2002. *Collingwood and the Metaphysics of Experience*. London/New York: Routledge.

D'Oro, G., and C. Connelly, eds. 2016. Collingwood and Philosophical Methodology. Special Issue of *Collingwood and British Idealism Studies*.

D'Oro, G., and S. Overgaard, eds. 2017. *The Cambridge Companion to Philosophical Methodology*. Cambridge: Cambridge University Press.

Donagan, A. 1962. *The Later Philosophy of R. G. Collingwood*. Oxford: Clarendon Press.

Haug, M.C., ed. 2013. *Philosophical Methodology: the Armchair or the Laboratory?* Abingdon/New York: Routledge.

Jackson, F. 1998. *From Metaphysics to Ethics*. Oxford: Oxford University Press.

Overgaard, S., P. Gilbert, and S. Burwood. 2013. *An Introduction to Metaphilosophy*. Cambridge: Cambridge University Press.

Rotenstreich, N. 1972. Metaphysics and Historicism. In *Critical Essays on the Philosophy of R. G. Collingwood*, ed. M. Krausz, 179–200. Oxford: Clarendon Press.

Toulmin, S. 1972. Conceptual Change and the Problem of Relativity. In *Critical Essays on the Philosophy of R. G. Collingwood*, ed. M. Krausz, 201–221. Oxford: Clarendon Press.

Williamson, T. 2007. *The Philosophy of Philosophy*. Oxford: Blackwell.

CHAPTER 2

An Essay on Collingwood

Bernard Williams

1 Three Reasons for Talking About Collingwood

My first reason for discussing R.G. Collingwood is to right, in a small way, a genuine injustice, and a disservice to the history of Oxford philosophy, which consists in the virtual obliteration of him from the collective local consciousness. In a book which appeared in 1958 entitled *English Philosophy since 1900*,[1] there is no reference to Collingwood at all.

Second, he differed in his whole approach to philosophy from his contemporaries: the pupils of Cook Wilson (as he was himself) and others, whom Collingwood called "the realists", such as Prichard, Joseph, and in Cambridge, Moore. He represented himself in his *Autobiography* (1939) as very isolated from them, and indeed rejected by them. It has been suggested by David Boucher, in a useful biographical essay,[2] that Collingwood somewhat exaggerated this. Certainly, any isolation can only have been increased by the *Autobiography* itself, which is a fascinating and often brilliant book—"strangely conceited, but instructive", Santayana said of it—but was written under conditions of stress and

Reprinted with the kind permission of Princeton University Press

[341] [Numbers in square brackets refer to the chapter's original pagination—eds.]

K. Dharamsi et al. (eds.), *Collingwood on Philosophical Methodology*, Philosophers in Depth,
https://doi.org/10.1007/978-3-030-02432-1_2

bitterly attacked the realists, in particular suggesting in its closing paragraph that "the minute philosophers of my youth, for all their profession of a purely scientific detachment from practical affairs, were the propagandists of a coming Fascism".

His intellectual connections were notably with Italy, particularly through his friend Guido de Ruggiero. There is historical disagreement how far he was influenced by Croce and Gentile (though he certainly disowned the latter as a worthwhile philosopher when he became a Fascist). He shared with them (significantly) an influence of Vico. Much of his special approach lay in his sense of history. He was himself a historian, and unusually—today, unthinkably—worked professionally in both philosophy and ancient history.

Born in 1889, he went to University College and took Mods and Greats, graduating in 1912. He became the Philosophy Fellow at Pembroke [342] College, and in 1927 he was appointed University Lecturer in Philosophy and Roman History. When he had written three books in ancient history, culminating in *Roman Britain* (1932) [in fact his third book was *Roman Britain and the English Settlements* (with J.N.L. Myres) (1936)—eds.], he was appointed Waynflete Professor of Metaphysical Philosophy in 1935. From this deep historical experience, he formed significant views on the philosophy of history; the history of philosophy; and the interpenetration, as he supposed, of the history of thought, and the study of metaphysics. All of this was unusual. The philosophy of history, because practically no one studied it. The history of philosophy, because practically no one really did it, although they said that they did. And the interpenetration of the history of thought and metaphysics, because it was entirely original. No one in Oxford and few in Britain and America have been so impressed by the significance of history as Collingwood, except for Berlin, who has acknowledged Collingwood's influence; and it is significant that Berlin, whose conception of what philosophy is, was rather influenced by positivism, gave up philosophy (at least as he himself identified the subject) in favour of the history of ideas.

The third reason for considering Collingwood is that he had original and some good ideas. He also had some bad ones, and many of his interesting ideas tend to be expressed in a confusing way. Partly this was because much of his significant work was produced towards the very end of his life when he was ill. He resigned from his chair in 1941 because of failing health, having had a breakdown in 1932 and repeated strokes from 1938. He died in January 1943, aged 53. Leaving aside two early philosophical works, he published *An Essay on Philosophical Method* in 1933, of which he said that it was "my best book in matter; in style, I may call it my only book". But he said this in his *Autobiography*, which was published in 1939, when he had produced in addition only *The Principles of Art*, in 1938. The *Essay on Metaphysics* came out in 1940, and *The New Leviathan* in 1942; he left a lot of material, which was edited by T.M. Knox (who was thought by most people to be an unfortunate choice), and published as *The Idea of Nature* (1945) and *The Idea of History* (1946).

2 Style

However, the problems of Collingwood's philosophy for those with a concern for sound analytical book-keeping are not only due to the circumstances in which some of it was written. Collingwood's contempt—and contempt was a sentiment which he quite often expressed—for the "minute philosophers" meant that he simply did not try to write philosophy [343] in the way that was favoured by them (and is still favoured by most of us some of the time and some of us all of the time), a way which aims to head off any ambiguity or any implication which some reader, perhaps a very perverse reader, might improperly take up: a style that is to say, which seeks precision by total mind control, through issuing continuous and rigid interpretive directions.

The *Essay on Philosophical Method* is in fact mostly a dull and dated book, full of what are likely to seem now unhelpful distinctions and assimilations, but it does contain a marvellous section at the end about philosophical style, called "Philosophy as a Branch of Literature".

> Every piece of philosophical writing is primarily addressed by the writer to himself. Its purpose is not to select from among his thoughts those of which he is certain and to express those, but the very opposite: to fasten upon the difficulties and obscurities in which he finds himself involved, and try, if not to solve or remove them, at least to understand them better.
>
> ...The Philosophers who have had the deepest instinct for style have repeatedly shrunk from adopting the form of a lecture or instructive address, and chosen instead that of a dialogue...or a meditation...or a dialectical process where the initial position is modified again and again as difficulties in it come to light.
>
> The prose-writer's art is an art that must conceal itself and produce not a jewel that is looked at for its own beauty but a crystal in whose depth the thought can be seen without distortion or confusion; and the philosophical

> writer in especial follows the trade not of a jeweller but of a lens-grinder. He must never use the metaphors or imagery in such a way that they attract to themselves the attention due to his thought; if he does that, he is writing not prose but, whether well or ill, poetry; but he must avoid this not by rejecting all use of metaphors and imagery, but by using them, poetic things in themselves, in the domestication of prose: using them just so far as to reveal thought, and no further.[3]

3 The History of Philosophy

Collingwood thought that philosophical writing laid certain responsibilities on the reader—a kind of patience; a postponement, though not a refusal, of criticism; and imagination to grasp what the writer's problem [344] might be. Moreover, he had a general principle, which we shall come back to, that any assertion only made sense as an answer to a question, and that you could not understand what was being said by an author unless you understood—this did not imply, explicitly formulate—the question that he was trying to answer.

This had particular implications for the practice of the history of philosophy. Peter Strawson, talking about Paul Grice, quoted with approval a remark of Grice's, which he claimed all Oxford philosophers would agree with, that "we should treat great but dead philosophers as we treat great and living philosophers, as having something to say to *us*". Collingwood would have agreed with this, so long as it is not assumed that what the dead have to say to us is the same sort of thing as the living have to say to us. He would not have agreed, that is to say, with Ryle's frequent injunction to treat something written by Plato, for instance, as though it had come out in *Mind* last month. Collingwood insisted, correctly, that the questions being answered by Plato and Hobbes, for instance, were not the same, and that you literally could not understand them unless you understood this. The "realists" had very little sense of this; and some of the most instructive (as well as enjoyable) pages in the *Autobiography* are those in which he addresses their methods.

> so, in ethics, a Greek word like *dei* cannot be legitimately translated by using the word "ought", if that word carries with it the notion of what is sometimes called "moral obligation". Was there any Greek word or phrase to express that notion? The "realists" said there was; but they stultified themselves by adding that the "theories of moral obligation" expounded by Greek writers differed from modern theories such as Kant's about the same

> thing. How did they know that the Greek and Kantian theories were about the same thing? Oh, because *dei* (or whatever word it was) is the Greek for "ought".
>
> It was like having a nightmare about a man who had got it into his head that *triêrês* was the Greek for "steamer"...[4]

If Collingwood had not been so ignored, it might not be necessary, as I am afraid it is, to remind people of this now. It might also not have been necessary to rediscover for oneself that the point of reading philosophers of the past is to find in them something different from the present—and that is not just a historical but a philosophical discovery. That much is true, even if we do not share the view which Collingwood evidently held, that there are no permanent problems of philosophy at all. It is not immediately clear why he held this, and the question is related to some [345] matters we shall come to later. It is worth saying now that it would follow from something that Collingwood did favour, a historicist application of a kind of holism, which may come from Vico, a tendency to think of an age as having an overall character.

Related to that, we do not, in order to sympathise with Collingwood's admirable outlook on the history of philosophy, have to share his very radical view, which he came to hold at the end of his life, about the historical character of metaphysics, something that I shall come to in the last section of these remarks.

4 Recovering the Past

The "realists" or "minute philosophers", having misidentified Plato's or Leibniz's questions, could announce that the philosopher had answered it wrongly. It was less obvious on Collingwood's own view what it was for Plato or Leibniz to have answered their own questions rightly or wrongly. In the *Autobiography*[5] he ties himself into something of a knot about this:

> If Leibniz was so confused in his mind as to make a complete mess of the job of solving his problem, he was bound at the same time to mix up his tracks so completely that no reader could see quite clearly what his problems had been. For one and the same passage states his solution and serves as evidence of what the problem was. The fact that we can identify his problem is proof that he has solved it: for we can only know what the problem was by arguing back from the solution.

On the same page, he says that this does not mean that the question "was Plato right to think as he did on such and such a question?" must be left unanswered. This would be as "lunatic" as to suppose that it was part of history that he beat the Corinthians by doing so. There is more than one thing questionable about this analogy; but the present point is that Collingwood, in deploying the analogy, seems to leave himself and us in the position that there can be an answer to the question "was Plato right to think as he did on such and such a question?", but only if the answer is "yes".

Now we could certainly deal with this by nudging Collingwood into a more reasonable position (though, remembering Collingwood's remarks about the dialectical character of philosophical writing, we should probably [346] say: more reasonable than the position he has reached at this stage of the argument). We do not have to assume that the philosophical writer has "made a complete mess" of the job. Nor need we assume that the only access to his problem is by arguing back from the solution. In fact, Collingwood would not disagree; and he did think that one relied on other parts of the text, and other texts, and other evidence, for the reconstruction of a question in the history of thought. So we are not locked in the paradox which we just glimpsed. But rather than simply taking on board these sensible modifications and qualifications, we should perhaps look at the question of why Collingwood might have been pushed in the direction of the paradox in the first place.

Here we need to address a broader doctrine that Collingwood held about historical understanding more generally. Probably Collingwood's most famous remark is "all history is the history of thought". History he held was "the history of *res gestae*, the actions of human beings done in the past".[6] The activity described in relation to the history of philosophy, of determining the writer's question and his answer to it, applies quite generally to understanding past action: to understand what an agent did, we have to understand what his question was, and what thought he had in answering it. The question "Why did Caesar cross the Rubicon?" is, if a properly historical question, to be taken as "What did Caesar think, as a result of which he crossed the Rubicon?". Moreover, there is no way in which Caesar's thought can be identified entirely from the outside; for the historian to answer the question, he has to have Caesar's thought—the very same thought—himself. "Thought can never be mere object. To know someone else's activity of thinking is possible only on the assumption that this same activity can be re-enacted in one's own mind. In that sense, to know 'what someone else is thinking' (or 'has thought') involves thinking it for oneself".[7]

This is the most famous of Collingwood's views, and the only one that I recall being discussed in the Oxford of the 1950s (except, very occasionally, some things that he had said about the philosophy of art, which I shall not discuss). There were several reasons for this being so. There was not much work being done on the philosophy of history, and Collingwood was very saliently one of the few recent and local contributors to it. Moreover, one of the few who was working on it, and the most senior, was W.H. Walsh, who had Idealist sympathies which inclined him to at least take Collingwood seriously. But most did not take him all that seriously, and the final reason for Collingwood being discussed at [347] that time was that the re-enactment thesis provided a convenient and seemingly rather dotty version of the kind of thing that a sensible empiricist-style account of history would want to avoid.

As Patrick Gardiner has said in a recent article,[8] a lot of this criticism was based on misunderstanding. One basis of the misunderstanding was to suppose that the re-enactment thesis proposed a method, but as Gardiner says, this is not the main point: "the question at issue is not so much one of the techniques the historian uses in his search after knowledge and understanding as one of what *constitutes* such knowledge and understanding when he has it."[9]

Another source of misunderstanding was a reading of the relation of thought and action in Collingwood in dualistic terms. This really was gratuitous: indeed, Collingwood emphasised the expression of thought in action, and in *The New Leviathan* he diagnosed dualistic errors in Descartes and was saluted for doing so by Ryle, Collingwood's successor as Wayneflete Professor, in his inaugural lecture.[10]

With the removal of these misunderstandings, we can be freed from the idea that what is at issue is some form of diachronic and self-confirming telepathy. Evidence is needed, and interpretation of action and circumstance helps the historian to a situation in which he can re-enact the thought.

Three other objections to Collingwood's account of history are worth rather more consideration.

Collingwood's typical insistence that re-enactment is involved in, and central to, *any* historical understanding, together with the thesis which that implies, that the subject matter of all history is individual action, unduly narrows the scope of history. This criticism seems to me unanswerable. Collingwood's view goes with a dislike of causes as opposed to reasons, something that is expressed, for instance, in his deeply eccentric attack on Thucydides for having turned history (which Herodotus had [348] initiated) into something different, namely social science.[11] It will be enough to say that history potentially involves, and often puts into the foreground, the understanding of actions in such terms.

A second objection is that the thesis may be thought to be, to use one of Collingwood's words, "lunatic": on his account, it seems, the historian in his moment of understanding Caesar, will be trying to cross the Rubicon, or doing the nearest thing he can do to that. But Collingwood had an answer to that, an answer which, he claimed, took him a great deal of trouble to get right.[12] This was the idea of "encapsulation", which depends on the point that the context in which the historian re-enacts the past thought—the context of the "real life" of the historian—is quite different from, and conflicts with, the presuppositions of that thought being appropriate. Having the thought, the re-enactment, is simulation, thinking in someone else's person. The content of the thought, if the attempt has been successful, is the same as that of the past agent's thought, but it occurs in the biography of the historian only in quotes or, as it were, indented.

There are two things that we should *not* say here. One is that the content is not asserted but entertained. The reason for not saying this is that the distinctions between assertion, entertaining, questioning and so on *recur* within the re-enactment itself. Nor should we say that the historian is "pretending" to think the thought. Pretending to x implies that one is not x-ing (or, at the very least, that one is not intentionally x-ing).[13] This is why not all dramatic acting is pretending; and it is very relevant, in the context of the present question, to think of dramatic acting which consists of improvisation.

A useful comparison is with another application of *Verstehen*, the ethnographer's stance. The ethnographer in the field can come to think as the people think with whom he is living, and he can enact their judgements and deliberations in his own person. But at the end of the line, these are not *his* thoughts. The difference from the historian is that the ethnographer may *lose* the end of the line: he may find it hard to come back to his regular life and may, as it used to be said, "go native". The historian cannot do that, unless he goes mad; the end of the line, at [349] which those thoughts which he thinks are not his thoughts, is all the time immediately in front of him.

The third problem with Collingwood's position concerns irrational or mistaken thoughts. William Dray, a philosopher who has done a lot to keep Collingwood's approach alive in the philosophy of history, referring to "empathy" theories in history, has said "one cannot re-think a practical argument one knows to be invalid".[14] In one sense, this seems to be wrong, and it can be met by appealing again to encapsulation. I can move in and out of an enactment, just enough to control it with my own sense of reality; we may compare, in the ethnographic case, the complex ambivalences of a modern investigator's approach to traditional magic. However, this only applies when I have identified the "invalid practical argument". It is a different question if I cannot get to the point of recognising anything as the practical question or reasoning which was the agent's; and this parallels, of course, the problem that we have already met in discussing the history of thought in the narrower sense, the problem of interpreting a radically confused text.

One straightforward thing to say is that there simply is a problem about interpreting a radically confused text, action, or set of actions. However, the problem is not as bad as it sounds, because there is a gradation from the intelligible, through things that are intelligibly confused, to that which is unintelligible. That last judgement does not simply return us to a situation in which, paradoxically, we are not in a position to make that very judgement. Precisely because there are such gradations, we can understand that something is unintelligible, without having to understand it.

However, this applies only to (relatively) particular cases—at most certain *sorts* of texts, actions, and persons. It cannot apply generally—or rather, we should say, in order to keep open a significant area of discussion, *too* generally. We can understand a given text or action only in terms of a local set of conventions, possibilities, and significances. It is what it is, means what it does, because of what it is possible to mean in those historical and social circumstances. (This reminds us again that re-enactment is not a kind of intuitive telepathy: it is based on knowing a lot about how things went there.) In understanding those circumstances themselves, how things went there, we do not have the same possibilities of detecting error or irrationality because there is no further set of considerations which provide enough leverage to enable us to make the separation implicit in claiming *both* that we understand what is going on *and* that we see in the light of this understanding that it is confused or [350] irrational. At some level of generality—and we should stress, again, in order to keep that area of discussion open, that we do not yet know what level of generality—we get to a point at which it is a condition on our understanding of what is going on that it should make sense under our interpretation, in such a way that it does not come out as seriously confused or irrational. As Collingwood rather marvellously said: "We call them the Dark Ages, but all we mean is that we cannot see".[15]

Collingwood considered history rather than social understanding more generally, for instance ethnography, but *this* idea is of course familiar. It is the idea of what is sometimes called a principle of charity, but we should reject that name in relation to Collingwood, as for ourselves, because of its entirely inappropriate suggestions of condescension. But we should also be careful with the title which is sometimes now more favoured, "the principle of humanity",[16] because that too strongly suggests a universalism of rationality, applying to human beings as such. Collingwood, with his strongly historicist outlook, was very suspicious of such an idea. It must be true that Collingwood, or any other hermeneutical thinker (not that he used that expression, so far as I know), will need to make some universal assumptions about human beings in order to render the historicism, the historical varieties of experience, intelligible as varieties of human experience; but the essential question is, how determinate and specific those universal assumptions are taken to be.

At any rate, since we are dealing with Collingwood, let us label the principle "the principle of constructive re-enactment" (PCR). We can now see an answer to the question we noticed earlier, why Collingwood backed himself into the seeming paradox of there being only one answer to the question "Was Plato right to think what he thought on a certain question?", namely the answer "yes". The reason is, I suggest, that with regard to really great thinkers such as Plato and Leibniz, Collingwood took it that the PCR applied to them directly and not just to the background in which they wrote. Collingwood was disposed to think that such writers were the only philosophers worth reading anyway, or at least, the only ones worth reading for their own sakes, as opposed to their providing evidence for the belief system of their time. And to them, with a few sensible modifications and qualifications, Collingwood thought that the PCR applied directly.

It is important that there are at least two different—and in some degree, opposed—reasons why this might be so. One is that such writers [351] embody or express in a particular intense and deep way the general assumptions of a period or epoch. The other is that they create new assumptions, help to bring into being new canons of what it is for theory or practice to make sense.[17] It looks in a way as though Collingwood ought really to favour the first option, a description of Plato or Leibniz, in terms of expressing rather than creating assumptions. However, there is a problem in even stating this question because, on Collingwood's last and most intriguing views about the nature of philosophy, it is hard to say what such philosophers were doing at all, in particular to the extent that they were doing metaphysics. It is this question that I shall now turn to.

5 Metaphysics and Presuppositions

We need first to look back at the method of question and answer. Collingwood said[18] that Bacon, in his epigram to the effect that the scientist must "put Nature to the question", laid down the true theory of experimental science. It contained the idea that the enquirer must take the initiative, deciding what he wants to know, and formulating this in the form of a question, and "he must find the means of compelling nature to answer, devising tortures under which she can no longer hold her tongue". *Mutatis mutandis*, Collingwood thought that this also contained the true theory of historical method. Equally, the historian must be active, having a question to which he needs an answer. Passive history Collingwood calls "scissors and paste history", the mere assemblage, with some critical mutual adjustment, of what the records have happened to leave to us. In archaeology, it was mere rummaging around, trying to find something. As he interestingly says,[19]

> "Let us see what we can find out about this site" is no more a "question", as I understand that term, than are such pseudo-questions as "What is knowledge?", "What is duty?", "What is the *summum bonum*?", "What is art?". Like them, it is only a vague portmanteau phrase covering a multitude of possible questions, but not precisely expressing any of them.

In *The Idea of History* Collingwood illustrates the idea with the straightforward example of a detective story in the English mode, where [352] the Poirot-like figure gets the right answers because he asks the right questions. Collingwood took himself to have used such methods in his archaeology. (Indeed, critics of that work have said that Collingwood only too readily interpreted what he found as the answer to his question.) In the *Autobiography*, he gives a rather famous account of how, in the First World War, he developed the idea because he had to walk each day to work past the Albert Memorial, and came to ask himself whether, in finding it loathsome to an incomprehensible degree, he was simply failing to understand the world from which it came and in which it was admired. In that work,[20] further, he discussed these ideas in terms of what he called "a logic of question and answer", but there is no reason to follow him in this, particularly given the ill-advised and arrogant remarks he offered there about what is ordinarily called "logic".

What matters for the present question is the elaboration of his ideas about question and answer that Collingwood introduced in the *Essay on Metaphysics.*[21]

There, Collingwood lays down the following theses in a rather formal style. (It will be obvious, e.g. in relation to the first, that they have to be understood to some degree as legislating about the notions involved.)

1. Every statement is made in answer to a question.
2. Every question involves a presupposition.

This means that the presupposition makes it a pointful, answerable question; indeed, makes it a determinate question. (It will be obvious that the identity of a question is not determined by purely linguistic means: one can ask the same question in different linguistic ways, and a different question by using different words.)

3. The logical efficacy of a supposition does not depend on the truth of what is supposed, or even on its being thought true, but only on its being supposed.

A supposition has "logical efficacy" in relation to a question if it causes (as Collingwood incautiously puts it) a question to arise: for instance, "this mark on a stone is an intentional inscription" has logical efficacy in relation to the question "what does it mean?".

4. A presupposition is relative or absolute: relative if it stands to one question as presupposition and to another as its answer; absolute if, for all questions to which it is related, it stands only as presupposition and not as answer. [353]

From this it follows

5. Absolute presuppositions are not propositions, in the sense of what is stated.

(5) does not mean that absolute presuppositions cannot be formulated or verbally expressed. It may be, possibly, that they do not have to be verbally expressed, but they certainly can be. To deny this would be inconsistent with many things that Collingwood says, for instance with the question[22] "Is this presupposition relative or absolute?" What "proposition" means is something that can be responsibly claimed to be true on the basis of proof, demonstration, verification, empirical evidence, etc.,[23] that is impossible in relation to an absolute presupposition because, if it could be verified, it would be the answer to a question.

This is illustrated by Collingwood's discussion of Mill on the principle of causation or the uniformity of nature. Such a principle was an absolute presupposition of science as Mill understood it, but Mill took, in typical positivist fashion, as Collingwood put it, a supposition for a proposition and got himself into a false position by trying to demonstrate it.

But metaphysics is the science of absolute presuppositions. Metaphysics does not propound (as defensible truth claims, that is to say) absolute presuppositions any more than natural science does, and it could not do so. It does not propound a certain kind of proposition, to the effect that certain absolute presuppositions are in fact presupposed. Thus all metaphysical questions and propositions are *historical*, and metaphysics deals with a class of historical facts.[24]

A historical reminder is in place at this point. A.J. Ayer's *Language, Truth and Logic* had been published in 1936, and Collingwood took it and its dismissal of the old gang of "realists" seriously. When Collingwood heard Prichard and Joseph complaining about the book and saying that it should never have been published, he said "Gentlemen, this book will be read when your names are forgotten".[25] But he thought that positivism's "anti-metaphysics" was misconceived, and that it was based on such things as a childishly resentful attitude to ethics or religion, supported by a fear that metaphysics was "malicious" towards science (a phrase of Russell's, from *Mysticism and Logic*). This was a misunderstanding, even if it was explicable why the positivists should have taken metaphysics to be an enemy of science: on their misconception of [354] it, metaphysics would have been an enemy of science, though not in the way that they supposed.[26] If metaphysics is what Collingwood says it is, the "historical study" of absolute presuppositions, then there are no objections to its status.

But is this account acceptable, or even fully intelligible? The first objection is an obvious one: why *historical*? Doesn't *our* metaphysics address itself to *our* absolute presuppositions? In a sense, Collingwood does not mind about this. The present, he says, "is the recent past",[27] and the kind of "metaphysical analysis", as he calls it, which serves to disentangle absolute presuppositions is very much the same with regard to the past and the present. Indeed, this is what you would expect from the re-enactment doctrine: to say that thinking about our presuppositions is like thinking about other people's is less surprising when one recalls that understanding other people's thoughts in the past involves thinking them yourself in the present.[28] These suppositions, importantly, are always in flux. Our present presuppositions are a transmutation of the past, and they are in course of transmutation towards the future. There are always "internal strains" in the "constellation of absolute presuppositions" at any given time. Metaphysical analysis can present those presuppositions and, presumably, the tensions between them, and this activity can be applied in one's own case. We may suppose that some of such arguments might look like trying to establish one absolute presupposition at the expense of another; if that is so, then, in Collingwood's view, that will be simply appearance, or alternatively some absolute presupposition has turned into a relative one.

However, there remain problems about the content of absolute presuppositions. The historical context I have mentioned, of replying to positivism, only serves to underline that problem. Consider some sentence "S″ which might traditionally be taken to express some metaphysical claim, the content of which is unverifiable. It is not much of an answer to the positivist to claim that the following is a historical verifiable proposition: We (or someone else) presuppose absolutely the content of "S". To interpret this historical proposition we need to know what relation to what is involved in presupposing the content of "S", and if the content of "S" is *ex hypothesi* suspect, this looks fairly daunting. It is important that this problem remains even if one is less worried than the positivists were by, specifically, verification.

Collingwood does not offer any general guidance as to how the content of absolute presuppositions is to be verified. The one example [355] with which he takes particular trouble, "God exists", does not help very much. The existence of God is an absolute presupposition for thinking done by Christians,[29] and as such it cannot be proved. A thinker, addressing himself to "God exists" as a metaphysical proposition, will put in the "metaphysical rubric" which yields a proposition to the effect that we presuppose in our thinking that God exists. Collingwood continues,[30] "If 'God exists' means [*sic*] 'somebody believes that God exists' (which is what it must mean if it is a metaphysical proposition) it is capable of proof … historical proof". He then points out that an archetypal proof of God's existence in the Christian tradition is the Ontological Proof, which indeed derives "God exists" from somebody's believing that God exists.

This may be a spectacular *tour de force*, but if it achieves anything, it achieves it only for this very peculiar case. Indeed, Collingwood's wider discussion of this case itself suggests that the content of absolute propositions [editors' note: the term Collingwood uses is "absolute presuppositions" to indicate that they have no truth value because the notion of verifiability does not apply to them; see also footnote 28], in its relation to the thinking of which they are the absolute presuppositions, can diverge a great way from the apparent semantic content of an absolute presupposition as verbally expressed. The presupposition that God exists is expressed, for instance, in our treating nature in our science as a unitary system.

So, two problems remain: that of identifying the content of absolute presuppositions, and of specifying the relation that they have to the thinking or activities of enquiry and so forth of which they are the presuppositions. Those two problems come together in the question already mentioned, of how an agent or group of agents are supposed to be related to what, when it is said that they presuppose a given absolute presupposition.

6 Collingwood and Wittgenstein

I think that this represents a genuine problem. It is the version in Collingwood's philosophy of a very real problem that presents itself to any philosophy which, like his, is hermeneutical and non-foundationalist. The problem concerns the relation between the discursive formulation of ideas or assumptions, on the one hand, and, on the other, practices and the description of practices. This is part of the question of how a *reflective* account of a practice, in particular of a practice of enquiry, can be given. Collingwood does not think that people can typically state the absolute presuppositions of their activities, and he makes a special point of saying that an absolute presupposition cannot be discovered by [356] introspection.[31] He does not think that people engaged in a practice necessarily or typically can be best described as *believing* the absolute presuppositions of their activities; the discursive formulation in which something is presented as an absolute presupposition and looks superficially like something that could be believed or asserted is a particular representation of an understanding which is implicit in and recovered from reflection on that practice.

Collingwood often speaks in terms of practices and changes of practices in a revealingly holistic way. In an interesting discussion of progress, he discusses a change in fishing methods between generations. The older generation will think that the old method is better than the new "not out of irrational prejudice, but because the way of life which it knows and values is built round the old method, which is therefore certain to have social and religious associations that express the intimacy of its connection with this way of life as a whole."[32] In *An Essay on Metaphysics*,[33] he speaks of a certain "natural piety", in Wordsworth's phrase, which could help the metaphysician simply to recognise facts—things simple and familiar, visible to the eyes of a child, and perhaps hidden from clever men because they are too clever. He saluted Samuel Alexander for expressing in an unassuming way such an outlook. Such things are not matters of proof or argument, not even of belief. The idea that the world is responsive to magic, for instance, is not a proposition which is refuted: "As long as you believe in a world of magic, that is the kind of world in which you live".[34] (And this can be taken as suggesting what, if anything, might count as "a belief in" a world of magic.)

Since absolute presuppositions structure practices; since they are not propositions in the sense of things believed; and since Collingwood's examples, though not his own account, suggest a difficulty in identifying their discursive formulations; then perhaps Collingwood should have acknowledged more than he did in this connection what in relation to Wittgenstein is called the primacy of practice, and the strength of Goethe's phrase, quoted more than once by Wittgenstein: *Im Anfang war die Tat*, in the beginning was the deed. Other resemblances to Wittgenstein, with regard to meaning as use, and the necessary expression of thought in action, have been noted by various writers.[35]

I should like to end not by suggesting a *compte rendu* of relations between the two thinkers but by mentioning one or two considerations [357] which Collingwood encourages us to bear in mind in these connections, more perhaps than Wittgenstein does.

(a) The account of metaphysics helps to bring out, though it does not solve, the question of the relations of reflection, and hence of philosophical reflection, to practice, and the status of discursive formulations of assumptions recovered from practice. Some Wittgensteinians seem to think that the primacy of practice means the primacy of descriptions of practice, and this, certainly, is a mistake, one that Wittgenstein himself did not make. However, for more than one reason (including, perhaps the "realist" origins of his own philosophical formulation and his ongoing fear that the activity of philosophy was itself pathological), Wittgenstein did not spend much time on considering how reflective descriptions of practice might be related to practice itself.

(b) Wittgenstein largely ignored history and disliked science. Collingwood respected science and based his entire philosophy on history. This at once places science as one activity among others, and reminds us of genuine and pervasive cultural variation. It therefore gives us richer resources for combating a stupid scienticism in philosophy since it invites one to think about the cultural role of science among other forms of understanding. At the same time, it provides a concrete sense of variations between actual "forms of life". The emphasis on history, again, also emphasises (a), since it adds to our reflections a sense of that variation.

(c) Collingwood's emphasis on historical change and constant tension in absolute presuppositions counteracts a danger which is implicit in talking in holistic terms about ways of life, the danger of falling into assumptions of functional coherence. Those assumptions simply install at the general cultural level Collingwood's paradox of the necessary rationality of a correctly interpreted text. I said before that Collingwood applied his principle of constructive re-enactment to whole systems and very great writers, rather than to every text or action taken particularly. However, while this is true, it is qualified by the emphasis on change. One has to make not just a system but a movement between various stages of systems intelligible, and one thing that makes this possible is that in its earlier stage the system of thoughts, understandings, or practices was in fact not fully coherent but was under tension.[36] Wittgensteinian accounts of social understanding have, notoriously, tended to favour a static picture of a fully functioning and coherent system.[37] [358]

(d) There is a well-known problem to be found in writers about these questions, particularly in Wittgenstein, which lies in their evasive use of "we", which may be taken in an inclusive sense, as implying universalistic preconditions on interpretation and intelligibility; or, in a contrastive sense, under which "we", here and now, are distinct from others elsewhere and elsewhen, who lived in other and different intelligible human formations. Collingwood has a good deal to offer on this set of questions. He himself tended to a radical historicism which played down, as I have said, the idea of permanent philosophical problems, and does not make much of assumptions which may be necessary in order to identify any way of life as a human way of life at all. He is helped, however, here once again by his emphasis on history, where continuity and change permit a developmental, diachronic understanding which is not offered by the blankly ethnographic case—still less by the purely imaginary and schematic ethnographic case which is favoured by Wittgenstein, the relation of which to "us" tends to be thoroughly ambiguous.

To consider, as Collingwood did, the case of history, and in particular the history of the Western world, is, of course, not a substitute for considering also the hermeneutical problems raised by ethnographic and similar studies. But it is, in itself, a deeply important case. Moreover, it is the case which is implicated very deeply in our attempts to understand the ethnographic case as well, and in our attempts to determine the relations between cultural variation and some universal assumptions about human nature (assumptions which, as we have seen, are necessary if we are going to be able to understand the notion of cultural variation itself). For it is a lively question, how far "our" understanding (and that is meant to preserve every ambiguity of "we") of the ethnographically exotic, of cultural variation, and hence of the humanly universal, itself indelibly carries with it assumptions drawn from our cultural experience as historically understood.

No adequate philosophical enquiry into interpretation can eventually ignore these questions, and so it must itself be informed by a historical sense. Collingwood reminds us of that fact; he reminds us of the problem and its importance; and, besides offering philosophical writing which in twentieth-century British philosophy is unrivalled in its brilliance, gives us fruitful lines of enquiry into that problem.

Notes

1. G.J. Warnock, *English Philosophy since 1900*, Oxford University Press, 1958.
2. 'The Life, Times, and Legacy of R.G. Collingwood' in *Philosophy, History and Civilization: Interdisciplinary Perspectives on R.G. Collingwood*, eds. D. Boucher, J. Connelly & T. Modood, University of Wales, 1995.
3. EPM: 209–14.
4. A: 63 following. The whole chapter is very worthwhile.
5. A: 69–70.

6. IH: 9.
7. IH: 288; compare A: 112 (about Nelson).
8. 'Interpretation in History: Collingwood and Historical Understanding' in Anthony O'Hear, ed., *Verstehen and Human Understanding* (Royal Institute of Philosophy Supplement 41, Cambridge UP, 1996), 109–119.
9. Page 114, Gardiner's emphasis. There are passages in which it is clear that this is what Collingwood means, although his formulation is misleading: for example, IH: 282 "what kind of knowledge has [the historian]?: in other words, what must the historian do in order that he may know [his facts]?"—A similar distinction between method and constitution may be relevant to contemporary discussions in the philosophy of mind (not primarily concerned with understanding the past) about the supposed distinction between "simulation theory" and "theory theory".
10. 'Philosophical Arguments,' in Gilbert Ryle, *Collected Papers*, Vol. 2 (London: Hutchinson; New York: Barnes & Noble, 1971), chap. 14.
11. IH: 28–31. The same bias comes out in the very poorly considered attack on psychology in part II of *Essay on Metaphysics*, which apart from anything else offends against Collingwood's own canons of how to read a text.
12. A: 112.
13. In this I side, in a famous disagreement, with Anscombe and against Austin. J.L. Austin, 'Pretending', and G.E.M. Anscombe, 'Pretending', *Proceedings of the Aristotelian Society*, Supplementary Volume 32 (1958), now in J.L. Austin *Philosophical Papers*, and G.E.M. Anscombe, *The Collected Papers of G.E.M. Anscombe*, vol.2, respectively.
14. *Philosophy and History: A Symposium*, ed. Sidney Hook (New York, 1963), page 113; quoted by Patrick Gardiner, page 118.
15. IH: 218.
16. The title is owed to Robin Grandy.
17. If one has an attachment to the idea of genius (as I think Collingwood did not, particularly), one can see this distinction as corresponding either to two theories of genius, or to two types of genius.
18. IH: 269.
19. A: 122.
20. A: 33, 42.
21. EM: 23 seq.
22. Allowed at EM: 40.
23. This point is made by Rex Martin in an article, 'Collingwood's Claim That Metaphysics is a Historical Discipline', in *Philosophy, History and Civilization*, at page 207.
24. EM: 49, 61–2.
25. Quoted by Boucher, op. cit. note 36.
26. There is a sparkling paragraph on this theme at EM: 169.

27. EM: 70.
28. This point does not entail the view, rejected by Martin, op. cit. page 216, that recovering or coming to know an absolute proposition is just an exercise of re-enactment.
29. EM: 186.
30. EM: 188.
31. EM: 188.
32. IH: 325.
33. Chapter XVII, 172 seq.
34. EM: 194.
35. See Boucher, op. cit. note 42 for references.
36. This is of course a fundamental Hegelian principle. Beyond this important point, I do not suppose that Collingwood was a Hegelian thinker, but to argue this would require a longer treatment, in particular of Collingwood's understanding of progress, and the extent to which he was in some sense, like Hegel, a relativist.
37. Peter Winch's *The Idea of a Social Science and its Relation to Philosophy* (London: Routledge & Regan Paul, 1958) was an early and influential example of this tendency.

CHAPTER 3

The Development of Collingwood's Metaphilosophical Views

James Connelly

1 Introduction

Louis Mink commented that: "It is evident from all of his work that Collingwood was an unusually self-conscious philosopher, preoccupied to an extraordinary degree with the problem of analysing his own thought in the way in which he had learned to interpret the thought of classical philosophers" (Mink 1969: 60). Hence for him, philosophy always self-consciously included the philosophy of philosophy. What follows is a short chronological guide to Collingwood's writings on philosophical method. It shows that there was a clear development of purpose and doctrine and that his metaphilosophical concerns arose partly out of the change in his own philosophical affiliations and his response to others holding differing conceptions of philosophy. For Collingwood, philosophical method was closely aligned to his identification of different domains of inquiry or forms of experience. These are not grounded in empirical features of the world but in how we approach each and any feature of the world. If one identifies different domains of inquiry the question of the methods appropriate to each arises; in the case of philosophy, as the methodological

J. Connelly (✉)
Hull University, Hull, UK
e-mail: j.connelly@hull.ac.uk

K. Dharamsi et al. (eds.), *Collingwood on Philosophical Methodology*, Philosophers in Depth,
https://doi.org/10.1007/978-3-030-02432-1_3

moment of all inquiry, this form of analysis generates conclusions specifically concerning self-reference.[1]

2 Reflecting on Realism

It is not always recognized that Collingwood was not born an idealist, nor did he have it thrust upon him; he achieved his idiosyncratic brand of idealism and in so doing espoused a distinctive view of his own. He was trained as an Oxford realist as an undergraduate (1908–12). It is important to recall that, far from idealism being the sole or dominant philosophy in the university, the dominant figures in his undergraduate years were ardent realists John Cook Wilson, H.A. Prichard, and H.W.B. Joseph, with others (such as W.D. Ross) becoming more influential, especially in ethics, in later years. Collingwood's philosophy tutor was E.F. Carritt, a realist; he attended Cook Wilson's class in logic and metaphysics and an advanced class in metaphysics. Collingwood greatly admired Cook Wilson and was great friends with Carritt, with whom he shared a lifelong interest both in aesthetics and in the philosophy of Benedetto Croce and the other Italian neo-idealists. However, as he broke with realism and developed his own philosophical approach, he began sharply to disagree with Carritt, Ross, and Prichard (and their Cambridge contemporary, G.E. Moore) over their conception of philosophy and its method.

Collingwood's philosophical career was interrupted by the First World War. However, he was able to complete and publish *Religion and Philosophy*, which was in part the outcome of his participation in the Cumnor group of modernist theologians led by B.H. Streeter and Lily Dougall, to whom he had been 'consulting' philosopher. During the writing of this book he chafed against the limitations of realist philosophy and method and signs of his later thinking are already to be found there. But he was dissatisfied with the book precisely on this account—and noted that fact in later notes on the book and its companion essay on 'The Devil', as he worked out his new dialectical philosophy from 1917.

As Collingwood wrenched himself free of his realist training he took stock of the waning idealism he found around him and of the alternatives available. He was scathing about pragmatism and he found the indigenous idealism to be wanting in historical sense (among other defects) and greatly preferred what was emerging from Italy, in the form of treatises by Benedetto Croce, Giovanni Gentile, and his great friend Guido de Ruggiero. As for the realists, Collingwood found their claim that knowing makes no difference to what is known to be a mantra without meaning and at best an unargued presupposition.[2] In addition, realism was guilty of taking all claims in the theory of knowledge to apply indifferently to each and every subject matter; hence realists found it unnecessary to accommodate different forms of knowledge. The result of this was not neutrality between different forms of knowledge, but scientism: natural science was taken as the paradigm of knowing and all other knowledge claims were evaluated by scientific criteria and forms of classification and, where differing in approach or outlook, found wanting. Collingwood's interest in the distinctive character of religion, history, art, and ethics was hence under siege. Secondly, the realists (and later the analytic school) were unwilling to give a full and proper account of their methods and presuppositions. In part it was precisely their unwillingness to lay bare their own presuppositions that led Collingwood to perform the task for them and therefore also for himself.

Collingwood began to rebel against his realist training shortly after graduation in 1912. This rebellion was not complete until around 1917–18. However, as mentioned, an important early influence was his discovery of the Italians, Croce, Gentile, and De Ruggiero, partly independently and partly under the influence of J. A. Smith and E. F. Carritt, who, despite his philosophical differences, took an especial interest in Croce's aesthetics. In this context, the significance of Collingwood's first published philosophical work being a translation of Croce's book on the *Philosophy of Giambattista Vico* (1913) should not be overlooked. In particular, Vico's concern with a 'new science' which was essentially historical was directly in line with Collingwood's later insistence on the importance of capturing the thoughts and intentions of past and present actors. His view that philosophy and history concern understanding, rather than scientific explanation, found resonance with the famous passage from Vico's *The New Science*:

> in the night of thick darkness enveloping the earliest antiquity, so remote from ourselves, there shines the eternal and never-failing light of a truth beyond all question: that the world of civil society has certainly been made by man, and that its principles are therefore to be found within the modifications of our own human mind. Whoever reflects on this cannot but marvel that the philosophers should have bent all their energies to the study of the world of nature, which, since God made it, He alone knows; and that they should have neglected the study of the world of nations or civil world, which, since men had made it, men could hope to know. (Vico, §331)

This is, of course, Vico's principle of *verum et factum convertantur*: "the condition of being able to know anything truly, to understand it as opposed to merely perceiving it, is that the knower himself should have made it ... it follows from the *verum-factum* principle that history, which is emphatically something made by the human mind, is especially adapted to be an object of human knowledge" (IH: 64–5).[3] Not only is this a clear precursor to his later doctrine of re-enactment, but it also contains a clear sense that different domains of inquiry require different modes of understanding.

Shortly after translating Croce's book on Vico, Collingwood delivered the first series of lectures of his career on Aristotle's *De Anima*. In *De Anima* Collingwood found the principle of the scale of forms operative in the articulation of philosophical concepts. Although the logic of a scale of forms was not fully and firmly worked out until the early 1930s and published as *An Essay on Philosophical Method* in 1933, it long formed one pole of Collingwood's approach to philosophical method. The key is that it rests on a recognition that philosophical thinking behaves differently to thinking in the natural, exact, or empirical sciences. Why does it do so? One answer is that "the philosopher's object is at once himself and his world, and hence philosophy and the philosophy of philosophy are identical. To philosophize is to face the question how we know; and this question is at the same time the question how we know that we know" (Collingwood 1925: 165). The key to philosophical method lies, initially, in reflexivity: philosophy has to account both for itself and for its relation to its object.

Collingwood's first major philosophical book was *Religion and Philosophy* (1916). At this point Collingwood was on the cusp of working through the realism in which he had been trained. Although he was moving towards a form of dialectical idealism, and this had methodological implications, there are many passages in which he is stating lifelong-held beliefs about the appropriate method of philosophy. For example, we find that it is inappropriate to start with a definition of religion because "No science is really in a position to define its subject-matter until it has brought its discoveries to a close" (RP: 3); the same argument is found in the opening pages of his last book, *The New Leviathan* (1942). Philosophy, especially, cannot proceed by stipulative definition laid down at the beginning of inquiry: any 'definition' it achieves is coextensive with its entire treatment of the concept under examination. This view was developed at length in *An Essay on Philosophical Method* and was ever present in Collingwood's philosophical thought. In *Religion and Philosophy* we sometimes find doctrines which Collingwood held throughout this philosophical life, but distorted by the realist framework in which they were embedded. In particular, he saw himself as having run foul of the static logic of realism which encouraged either 'lumping' or 'splitting': that is, either asserting (on the basis of similarity or overlap) the identity of concepts or, on the other hand, taking distinctions to be sheer differences. As he came to see, realist logic was unable to account for the movement of philosophical thought in which both identity and difference are mutually necessary, concepts overlap, distinctions do not require splitting concepts asunder, and similarities do not require asserting their identity. In short, he was developing a dynamic conception of philosophy and its method and repudiating the static concept he found in his erstwhile realism.[4] He later wrote of this development in *Speculum Mentis* (1924) and in notes on his philosophical development. In *Religion and Philosophy* Collingwood had not adopted a dialectical approach and his procedure betrayed tendencies which he later classified as errors. For example, in the earlier work, he identified history and philosophy (RP: 51), thereby committing the 'fallacy of identified coincidents'. In writing *Speculum Mentis* (an explicitly dialectical work) a few years later he reconsidered their relationship and drew an important distinction between implicit and explicit features of experience. There he wrote that:

> With much of what [*Religion and Philosophy*] contains I am still in agreement; but there are certain principles which I then overlooked or denied, in the light of which many of its faults can be corrected. The chief of these principles is the distinction between implicit and explicit. I contended throughout that religion, theology, and philosophy were identical, and this I should now not so much withdraw as qualify by pointing out that the 'empirical' (i.e., real but unexplained) difference between them is that theology makes explicit what in religion as such is always implicit, and so with philosophy and theology. This error led me into a too intellectualistic or abstract attitude towards religion, of which many critics rightly accused me. (SM: 108n)

Here we see Collingwood correcting an error which was the outcome of the brittle binary logic of realism; later, in *An Essay on Philosophical Method*, he developed a complete account of the relationship between concepts allowing both for their identity and their overlap. Between writing *Religion and Philosophy* and *Speculum Mentis*, therefore, Collingwood underwent a conversion to a form of dialectical idealism[5] in which method and object, in order to be adequate each to the other, had to be construed as activity, not as binary static oppositions.

In 1918 Collingwood wrote of *Religion and Philosophy* that it "represents the high-water mark of my earliest line of thought—dogmatic belief in New Realism in spite of an insight into its difficulties which I think none of my teachers shared ... The whole thing represents a point of view I should entirely repudiate" (Collingwood 1918). And in a similar note on 'The Devil' he wrote that it "represents the breaking point of my earlier philosophical beliefs. It is still realism, sharpened and hardened ... The flagrant superficiality of it ... drove me back upon my real convictions, & led to a year of negative criticism (1916a) & the building-up of a new dialectical idealism in 1917" (ibid.).

This "building-up of a new dialectical idealism" took the form of a full-length book entitled *Truth and Contradiction* (1917). This was offered to a publisher but never published. As only the second chapter survived, commentary on the whole book is impossible; however, in this chapter he analysed the strengths and weaknesses of the coherence theory of truth and rejected the associated notion of degrees of truth and reality. In this endeavour we can see the truth of his later comment that he was building up a new dialectical idealism, but at this stage had still not yet achieved it; *Truth and Contradiction* was his first serious, but not entirely successful, attempt. The flavour of the book is brought out by Henry Jones's comments made as reader for Macmillan:

> I do not know any writer more frank. He cares not one whit to what extent he exposes his flanks to his critics, and makes statements which, taken by themselves, look either purely absurd or preposterously untrue. But that is only one side: on the other is the fact that these statements are stages or steps in the development of his main argument, half-truths or sheer errors in which it is not possible to rest and which just compel a movement onwards to a wider truth. And this is precisely what he means, and rightly means, by 'dialectic', the way in which 'the development of thought' takes place ... By exposing the inadequacy of the other doctrines he means (and exemplifies) the discovery of the truth that the theories contain, and reconciling the half-truth each contains in a wider truth. This is a version of the 'dialectic' of Plato and of Hegel; and in a sense not new. But it is done in a fresh way: clear, frank, interesting, and some-how very 'taking' ... All the time he is showing

> the true nature of Philosophy, and finding that movement, activity, process is the living soul of all thinking and of all objects of thought. (Jones 1917)

Jones has clearly identified the theme that philosophy is dialectical thought and it is dialectical because its object is activity. This was emphasized in *Religion and Philosophy* in relation to philosophical (rather than psychological) thinking about the mind. Mind is what it does; there is no mind other than its activities, or to be precise, other than the activity of thinking: "The idea of the mind as a thing distinguishable from its own activities does not seem to be really tenable; the mind *is* what it *does*; it is not a thing what thinks, but a consciousness; not a thing that wills, but an activity" (Collingwood 1916b: 34). This is, of course, the formula derived from Gentile and De Ruggiero; it is the core of actualism and reveals the influences now working within Collingwood's thought. *Truth and Contradiction* develops this. It emphasizes activity; error as incomplete truth; and the idea of truth requiring the inclusion of its opposite within itself. In chapters and essays from 1919 onwards we can see Collingwood grappling with these themes and seeking appropriate language to express this satisfactorily in relation to other philosophical approaches of the time.[6] He chose not to rely on the old idealism of Bradley, Green, and Bosanquet, nor on Hegel; he was influenced by the Italians but took the view that simply regurgitating their views to an English audience would not convince it. He was searching for his own way of conceptualizing the issues and for his own arguments so as to engage the realists and the emerging analytic school in philosophy on their own terrain. No one else was doing this at the time.

3 The Practice and Method of Philosophy

Collingwood frequently took the opportunity to speak to audiences, whether archaeological or philosophical, outside the university. One such occasion was his address at the conference marking the centenary of John Ruskin's birth, held in Coniston in 1919. In this lecture he asked the question what it meant for Ruskin, who did not think of himself as such, to be thought of as a philosopher. This he does by drawing a distinction between philosophy as act and as fact (as Gentile might say).[7] Philosophy is hence both an activity and the beliefs people hold: the former is the activity of philosophizing, the latter is the possession of a 'philosophy'. A person's 'philosophy' is constituted by a 'ring of thought' and

> This central core of convictions ... exists in each one of us, and in that sense every one of us has a philosophy. But for the most part we do not know that we possess it: still less do we know what are the convictions which constitute it. The fact seems to be that a man's deepest convictions are precisely those which he never puts into words. Everything which he says and does is based upon his grasp of these convictions; but just because his grasp of them is so complete, so unquestioning, he never finds it necessary to express them at all. ... It may seem strange that our deepest and most important convictions should habitually go unexpressed. But this is still stranger, that we are often quite mistaken as to what these convictions are. If you ask a man to state his fundamental beliefs, and then carefully watch his actions and sayings, you will generally find that these are based on a set of beliefs quite different from the ones which he has stated ... Now it is this attempt to discover what people's philosophy is that marks the philosopher ... everybody has a philosophy, but only the philosopher makes it his business to probe into the mind and lay bare that recess in which the ultimate beliefs lie hidden. (Collingwood 1922: 7)

Here is something very like, 20 years *avant la lettre*, an account of the doctrine of absolute presuppositions expounded in *An Essay on Metaphysics* (1940).

In the early 1920s, at the time of his most focused interaction with Italian neo-idealism and the work of de Ruggiero in particular, Collingwood produced a stream of essays and articles on the nature of philosophy.[8] The common theme was that they all insisted that philosophy takes its object as activity and that its method has to accord with the nature of the object. In 'Croce's Philosophy of History', Collingwood objects to the elements of naturalism in Croce's philosophy. Croce, he claims, makes thought and the spirit into an object, not a subject. It becomes that which thinks, not the thinking itself. The problem is that

> there are two Croces, the realist, dualist, empiricist, or naturalist, who delights in formal distinctions and habitually works in dualistic or transcendent terms, and the idealist, whose whole life is a warfare upon transcendence and naturalism in all their forms, who sweeps away dualisms and reunites distinctions in a concrete or immanent unity. (Collingwood 1921: 267)

Croce, for Collingwood, cannot escape the yoke of naturalism with its sharp and brittle formal distinctions, and his failure to transcend naturalism results in an unsatisfactory philosophical method. For example, Croce sometimes maintains that history and philosophy are identical; sometimes he regards philosophy as subservient to or absorbed by history. Collingwood approves of the identity of philosophy and history, but he is careful to say that this cannot be a simple empirical identity or one based solely on the existence of common characteristics: identity is a dynamic conception; philosophy and history are united in every real act of thinking: "Each without the other is a lifeless corpse: every piece of real thinking is both at once. This is Gentile's view" (Collingwood 1921: 276). In the same year (1920) Collingwood wrote a sequence of experimental manuscripts, beginning with *Libellus de Generatione*. Collingwood states that this work

> is an attempt to sketch the outlines of a philosophy to which I attach the name of Absolute Empiricism. In giving it this title I intend to emphasize its connexion with the Empiricism of Hume; a connexion consisting in the absolute denial of any such concept as substance and the resolution of all reality into the actuality of experience. My fundamental doctrine is that reality is becoming, that is to say reality not so much *is* as *happens*, which implies that the reality of mind is the process of its experience, its life, and nothing else. Nor do I admit any dualism between mind and its object such that while mind is wholly process its object can be conceived as a static whole outside it. The object is process too, and these are not two processes but one process.[9] ... And yet the title of absolute empiricism contains a suggestion of what I have called 'subjectivism,' which I would willingly avoid. I find even the modern Italian idealists ... tinged with this subjectivism, which I regard as belonging to realism rather than to what I should like idealism to be. (Collingwood 1920a: 1)

The rest of the essay consists of working out and criticizing the implications of realism as a static conception of reality and the idealist alternative to it. *Libellus* was swiftly followed by essays developing key themes: 'Notes on formal logic', 'Notes on Hegel's Logic', and 'Sketch of a Logic of Becoming' in September, culminating in a 'Draft of opening chapters of a "Prolegomena to Logic" (or the like)' at the end of the year. In the 'Prolegomena' he argues that we have to turn away from empirical logic in which we are "thinking about thought as if thought were an object of thought like any other; [instead] we must now turn to the search for a concept of logic, ask by what right we practice this empirical logic; that is to say, we must turn to the pursuit of transcendental logic" (Collingwood 1920d: 26). Transcendental logic is required for thinking about objects where the object is the act of thinking itself. Logic is therefore its own object, it is activity, not a Platonic entity and it is self-referential and therefore has to account for itself:

> For logic, the characterization of its object as *scientific* thinking or *logical* thinking is meaningless. Its object is fully and finally characterized as *thinking* … In logic thought is and can only be of itself; therefore this is the only possible kind of thought. All thought must be thought of itself. This is proved by the existence of logic defined as the science of thought—the science in which the object of thinking is the activity of thinking. (ibid.: 29–30)

The problem, however, is that the logic we find around us "does not do this: the process of so-called thinking is quietly set aside by the logician, who concentrates his analysis on something quite different, not *thinking* but *thought*, not an activity of process but a state, indeed a perfectly quiescent state, of apprehension" (ibid.: 40). Here Collingwood is making a distinction between thought and thinking: the latter being the process, the activity, and the former the ossified result of thought. This is Gentile's distinction between *pensiero pensante* (the act of thinking) and *pensiero pensato* (thought as the product of thinking), or the distinction between act and fact; and it is the actualism or actual idealism of Gentile and de Ruggiero.

In the early 1920s, then, Collingwood was searching for the proper approach to understanding the nature of philosophy and its method and he had provisionally found it in the writings of the Italian neo-idealists. However, although he found himself in possession of a serviceable philosophical method (employed in his lectures and in *Speculum Mentis*), he was aware that it still lacked adequate philosophical justification. It took him a further ten years to work this out to his own satisfaction.

In 1923 Collingwood contributed to a symposium entitled 'Can the New Idealism Dispense with Mysticism?' Here he made clear the distinction between the implicit and the explicit and the role of philosophy in making explicit the implicit. He makes a careful distinction between the mediate and the immediate: all experience has implicit features and the role of the philosopher is to render them explicit, but by implicit is not meant immediate. He does not deny the reality of mystical experience, but he denies that mystical experience is immediate. He states that "Modern idealism maintains that all experience is mediate and therefore it is bound to show that mystical experience is mediate too, and that the traditional account of it as intuitive mutilates and distorts it" (Collingwood 1923a: 280). And he continues:

> the mediation which is actually present [in mysticism] is not wholly explicit: the mind reaches truths, but does not know how it has reached them. It may even think that it has not reached them by *any* path, that is by any describable process of thinking; but this, if it is believed, is wrongly believed. The truths in question are reached somehow, and it is the business of scientific or philosophical thought to lay bare this concealed process, to render explicit the mediation which in the mystical experience itself was only implicit. (ibid.: 281)

For Collingwood, then, the distinction within experience between mediate and immediate knowledge is one of degree. Knowledge is always mediate, but it is more or less consciously so, more or less self-critical, more or less aware of itself; philosophy draws out the implicit features of experience by making them explicit. Knowledge will sometimes appear to be immediate simply because, to the knower, the mental and logical processes on which that knowledge depends are beyond its grasp at that moment. His argument against the understanding of knowledge and experience provided by mysticism is akin to his argument against realism construed as direct or intuitive knowledge of an object (whether in perception or in ethics). He grants that, prior to analysis, that is how it might seem to the knower, but he does not accept this as the final word; the belief that one is in possession of direct or intuitive knowledge does not imply that one genuinely is. Realism ignores the mediacy of experience and hence, like mysticism mischaracterizes it.

In 1923 Oxford university became embroiled in a debate concerning the introduction of a new honours school of Philosophy and Science or 'Science Greats'. Debate swirled around the various institutions and committees of the university and featured in the *Oxford Magazine* where Collingwood made an appearance. His comments are instructive for helping us in understanding his approach to philosophy, especially by contrast with natural science. He began by arguing that

> Any problem which arises out of the development of scientific thought must be soluble, if at all, only by a further development of the same kind of thinking and a philosopher, with whatever admiration and interest he may watch the work of scientists, has no more right to forestall the result of their inquiries by an edict as to what is "philosophically admissible" than to tell the archaeologists what it is philosophically admissible for them to find in the inner chamber of Tutankhamen. (Collingwood 1923b: 301)

Science and philosophy, he continued, are different things, and "each demands a specialised training which differs from that required by the other ... the chief need of philosophical thought to-day is to escape from an undue sense of the importance to itself of scientific methods. Philosophy has since the Renaissance done much to impede its own progress by aping science, as science did in the middle ages by aping philosophy" (ibid.: 301). He went on to maintain that the sciences exhibit in various ways one form of thought, and share, with modifications, one technique; while philosophy is a

> different form of thought with a different technique, superficially similar in that science and philosophy both use the verbal form of the universal proposition, but profoundly dissimilar in the meaning that they attach to this verbal form and the process by which the meaning is reached. To say that science and philosophy are the same thing because they both *generalise* is, to my mind, as serious an equivocation as to say that fiction and history are the same thing because they both *narrate*. (Collingwood 1923c: 340)

This statement is a clear anticipation of *Speculum Mentis*, where he argued that scientific thinking is abstract thinking, concerned with abstract universals, whereas philosophical thinking is concrete thinking and leads to a concrete universal. Whereas the scientist "treats a phenomenon as a case of a law, an example of a type, and ignores everything in it which does not belong to it in this capacity", the philosopher "refuses to simplify his task by this preliminary abstraction, but insists on seeing the object in its entirety or not at all" (ibid.: 340). Further, science and philosophy have a different object:

> Science is based on hypotheses or assumptions, and in any science there are certain assumptions which the scientist as such does not allow himself to question. ... But the business of philosophy is precisely to question all assumptions wherever it finds them ... These radically divergent attitudes towards the question of hypotheses lead to a profound difference in the methods and structure of science and philosophy respectively. All the sciences work in one way, and are therefore connected by the possession of common methods and principles, and thus find ... little difficulty in exchanging problems. For they all agree in making abstractions, and only differ in making different abstractions. A scientist is therefore apt to approach philosophy by assuming that it too makes a specific abstraction, or ... has a specific "field." But philosophy makes no abstraction, has in that sense of

> the word no field: it is the attempt to overcome abstraction, to avoid the arbitrary cutting-up of reality into compartments. (ibid.: 340)

By the time of writing *Speculum Mentis* in the spring and summer of 1923 Collingwood had become confident that he had within his grasp an approach to philosophy which would enable him to elucidate the relationship between forms of experience, of which he distinguished art, religion, science, and history, with philosophy arising as critical reflection on the other forms of experience. The relations between art, religion, science, history, and philosophy are displayed as a scale of forms of experience in which each term in the scale renders explicit what for the previous term was only implicit. The scale of forms had by this time been identified (at least in his lectures on moral philosophy in 1923) as typical of how philosophical concepts articulate themselves, hence therefore as an appropriate guide to philosophical method.

From 1919 to 1940 Collingwood delivered lectures on moral philosophy annually. In Part II of the 1923 lectures on moral philosophy Collingwood for the first time offered an account of philosophical method featuring the idea of a scale of forms. In briefly outlining its main features he attributed the idea to Aristotle and cited as an example the analysis in *De Anima* of the different forms of the soul differing in function and capacity. These lectures were written immediately after *Speculum Mentis* where Collingwood had developed the distinction between implicit and explicit features of experience and philosophy as experience raised to self-consciousness, rendering explicit the principles implicitly informing experience and showing how each form gives rise to its successor. In his lectures he stated that

> Our series is to be a series of the forms of action; and action is the opposite of passivity. Hence at the bottom of the scale we ought to find pure passivity and at the top, pure activity. Every stage in the scale ought to be more active than the one below it and more passive than the one above it; and the dialectical process leading from one to the next must be based on the lower stage's incomplete self sufficiency, its dependence on a principle which it does not itself include or possess. (Collingwood 1923d: 11)

By 1923 Collingwood had, then, became a dialectical thinker. But the implications of this should not be misconstrued. Although he was inspired by Hegel and the Italian new idealists, we do not find him extolling the virtues of the Hegelian dialectic (other than by implication); we rarely hear the language of Hegelian/British idealism; and we certainly never find him, in developing his philosophical method, saddling up the tired old war horse of thesis, antithesis, and synthesis popularly supposed to be characteristic of Hegel's thought. In other words, he wanted his approach to be accepted (or rejected) on its own merits rather than on the basis of borrowed authority from other authors. He wrote in *Speculum Mentis*:

> I could easily have pointed out the affinities of my position with that of eminent writers past and present, and so, perhaps, have recommended it to readers who rightly shrink from any philosophy which is advertised as new. If I have consistently refrained from doing this it is only because I want my position to stand on its merits rather than on names of great men cited as witnesses for its defence. (SM: 12–13)

Progress on working out the dialectic of implicit and explicit and developing a satisfactory conception of the scale of forms accelerated from 1927–9 onwards, in his writing on moral philosophy, politics, and the philosophy of history. This required consideration of the nature and distinctive character of philosophical concepts. Following *Speculum Mentis* and prior to *An Essay on Philosophical Method* Collingwood made some important statements on method in articles on economics and politics and also in a lengthy prolegomenon to the philosophy of history.

In 'Economics as a Philosophical Science', Collingwood states that what he intends to do is to throw light "on some of the fundamental conceptions which economists do not so much derive inductively from facts as presuppose in their attitude toward the facts" (Collingwood 1926: 162). These conceptions, such as value and wealth, are used by economists but "seldom if ever satisfactorily defined". He proceeds to advance the thesis that they "are various aspects of, or various attempts to describe, a certain form of action which … we shall call economic action. The conceptions of value, wealth, and so forth are not ultimate inexplicables; they can be understood, but only by resolving them into the conception of economic action. This resolution is a task for philosophy" (ibid.: 162). He then draws a distinction, central to his mature methodology, between the empirical and philosophical sciences:

> Philosophical thought is that which conceives its object as activity; empirical thought is that which conceives its object as substance or thing. Economics, then, is an empirical science if it is conceived as the study of a thing called

> wealth; philosophical, if it is conceived as the study of economic action. But it is not enough, to make a science philosophical, that it should *call* its object action; it must really *think* of it as action. If a certain type of thought calls its object activity but nevertheless deals with it as if it were a substance or thing, it is an example of empirical, and not of philosophical, thought, and this is what happens in the case of psychology. Empirical economics, as belonging to the empirical study of human conduct, may be regarded as a branch of psychology: and just as there is a philosophical science of knowledge, distinct from the psychological science of knowledge, so there is a philosophical science of conduct (ethics) distinct from the psychological science of conduct. (ibid.: 163)

By 1927 Collingwood had introduced the idea that philosophy deals with universal and necessary features or experience: "When we speak of the philosophy of something (e.g. of art, of religion, of history) we mean to designate a body of thoughts which arise in us when we think about that thing. These thoughts must be philosophical; that is, they must be universal and necessary" (Collingwood 1927: 335). It is this which distinguishes it from scientific thought, which is universal "only in the sense that it is universally applicable to a limited sphere; it is empirically universal, not absolutely universal; it applies to all the facts that make up the field of an inquiry, but not to all facts whatever, on the contrary" (ibid.: 335). For Collingwood, a philosophical account of a form of experience is also an account of its place among various forms of experience. It is central to this view that forms of experience are forms of experience for everyone, not merely for those designated historians, artists, or scientists. History, or any other form of experience "is a transcendental conception … when regarded as a pure form of activity; though it becomes, like them, an empirical conception when it is arbitrarily restricted to certain specialized embodiments of that form" (ibid.: 357). The philosophy of history, for example, is then the "exposition of the transcendental concept of history, the study of history as a universal and necessary form of mental activity" (ibid.: 357). Another consequence is that requiring a definition of (in this case) history as a starting point would be mistaken because, by contrast with an empirical concept (which requires definition because it is neither universal nor necessary) a transcendental concept "need not be defined, because we are all possessed of it so far as we think at all; nor can it be defined, because, being necessary to all thought, it is necessarily presupposed in its own definition and the definition thus becomes circular … Definition, in short, is an operation peculiar to empirical conceptions" (ibid.: 357–8).

In 'Political Action', an article stressing the importance of conceiving both philosophical thought and its object of study as activity, Collingwood argued that political theory is typically conceived as the theory of the state regarded as a thing possessing certain attributes, with its task being to scrutinize this thing and provide an account of its attributes. He then remarks that

> A theory so conceived moves within the four walls of the category of substance and attribute. The state is a substance, an existing entity, and sovereignty, or whatever word one uses to indicate the sum total of its powers, is its essential attribute. A deductive political theory may work out, a priori, the implications of this essence; an inductive theory may collect information about the various attributes of sovereignty found to subsist in various states; but all such theories, deductive or inductive, are agreed in accepting the limitations of the category of substance and attribute. (Collingwood 1929a: 155)

He announces that he proposes to approach political theory differently by replacing the question 'What are the attributes of the state?' with the question 'What is political action?' In so doing he will move from the category of substance and attribute to the category of action. He states that his reason for making this proposal

> is not any lack of respect for the category of substance and attribute … nor any intention to disparage the theories of the state that are based upon it. Theories of the state, both empirical and a priori, are indispensable to any political theory … But all theories of the state … find themselves confronted, sooner or later, with certain questions that they cannot answer; and these questions are all such as concern the limits of the state, and its relations with other bodies, be they states, or churches, or trade unions, or municipalities. These are the very questions which, alone of all the questions in political theory, are today matters of urgent practical importance; and, with the example of the great political theorists before us, we need not apologize to each other for thinking primarily, in our philosophies, of the things that matter most in our practice. (ibid.: 155–6)

However, for Collingwood, these questions, although insoluble in terms of substance and attribute, are soluble in terms of action:

> if you think of politics as a thing, the state, or a number of things, the various different states, having this and that attribute, then the mere 'thing-

> hood' of the state, as so conceived, implies a rigidity of conception, an intransigence of behaviour, which gives rise to a simple destructive dilemma. If the state gives way to the trade unions, or signs the covenant of the League of Nations, either it is surrendering its sovereignty and therefore ceasing to be a state, or it is merely showing that it never possessed such a thing as sovereignty, and therefore ceasing to pretend to be a state. In either case, the state is a discredited superstition. (ibid.: 156)

Later in the same year Collingwood delivered the first fully rewritten series of lectures on moral philosophy since 1923, stating again, and firmly, that a "philosophical concept is universal in the sense that it arises necessarily whenever anybody thinks about a subject ... the concepts which compose the body of philosophy are transcendentals" (Collingwood 1929b: 12).

We have seen that in the years leading up to 1933 Collingwood, either in his introductions to his lectures or in his manuscripts or chapters on ethics and politics, was gradually working out and refining the themes later expounded in *An Essay on Philosophical Method* through examination of the distinctive nature of philosophical concepts versus non-philosophical or class concepts, and the implications of the resultant conceptual overlap for philosophical analysis. However, he had not yet fully worked out the detailed logic of the overlap of classes and his vocabulary had not yet fully settled. For example, by the time he wrote his methodological introduction to his lectures on moral philosophy in 1932 he had dropped the term 'transcendentals'. One reason for this might have been a growing recognition on his part that he needed to tailor his arguments to appeal to an audience, both student and professional, who might be more inclined to listen if he abandoned the old philosophical language and engaged the new. It is important to recall that between 1928 and 1931 Collingwood's Oxford colleagues had been engaged in an "informal symposium ... on the idea of Right in Morals" (Muirhead 1932: v). The symposiasts were W.D. Ross (*The Right and the Good*, 1930), E.F. Carritt (*The Theory of Morals*, 1930), H.W.B. Joseph (*Some Problems in Ethics*, 1931), J.L. Stocks, in articles published as *The Limits of Purpose* (1932), Muirhead himself (*Rule and End in Morals*, 1932). The terms of the symposium originated partly in H.A. Prichard's 'Does Moral Philosophy Rest on a Mistake' (1912), his inaugural lecture on 'Duty and Interest' (1928), and *Duty and Ignorance of Fact* (1932) and the groundframe was provided by G.E. Moore, whose books *Principia Ethica* (1903) and *Ethics* (1912) set much of the agenda.

Against this background, Collingwood's dialectical approach to the arrangement of concepts in moral philosophy (in particular his serial arrangement of utility, right, and duty) now had to face criticism and respond to these rival accounts from a rival philosophical standpoint which shared an opposed conception of philosophical method. In particular, it seems that Ross's *The Right and the Good* acted as a stimulus to Collingwood's thought on method, especially (but not only) in ethics. In his view, Ross's approach shared the weaknesses of Moore and Prichard and Carritt. Leaving aside specific points of analysis (e.g. their divergence in the analysis of right and duty) Collingwood and the others diverged considerably on the appropriate method of analysing concepts. In a chapter entitled 'Utility, Right and Duty' Collingwood discusses this and states that

> Utility, rightness and duty have been called three forms or kinds of goodness. This implies that goodness is a generic concept of which the other three kinds are specific determinations. Now, the traditional formal logic, which in this matter is still followed by those schools of thought which most emphatically call for its reformation, holds that a concept is specified by disjunction. If A is a concept and α β and γ its three specifications, then every example of A is an example of either α, or β, or γ; and every example of α, or of β, or of γ, is an example of A; but no example of α can be an example of β or γ. The three classes α, β, γ, exclude each other and together make up the class A. (Collingwood 1932a: 21–2)

This, he says, is an accurate description of the mathematical and empirical sciences, but it is not applicable to the specification of goodness into utility, rightness, and duty because

> it would mean that every action which is good is either useful, or right, or obligatory, and that, whichever of those three qualities it exemplified, there would be two others which it did not. If an act is useful, it cannot be right, and it cannot be a duty; if right, it cannot be either a duty or useful; if a duty, it cannot be either useful or right. These conclusions are all plainly false. (ibid.: 22)

His argument then hinges on the claim that, since it is possible for the same act to be useful, right, and a duty, the concepts are not related as species to genus in traditional logic, but otherwise. He commits himself to exploring this other way, and this, in its fully explicated form, comprises

An Essay on Philosophical Method. Meanwhile, he points out an important corollary of understanding relations between concepts in the manner he rejects:

> If concepts are related to one another by disjunction (so that any A is *either* α or β, but cannot be both), then one possible method of investigating their nature is to collect instances of A and *sort* them into α's and β's. When we wish to arrive at general conclusions about α, we can obtain them by examining the collections of α's, confident that, since the presence of α implies the absence of β, we are in no danger of noting peculiarities of β by mistake for peculiarities of α. But suppose α and β were so related that one and the same individual might fall into both classes. This method would become worthless. We could no longer say, when examining an instance of α, "whatever I see here must be something belonging—essentially or accidentally—to α and not to β," because what we are examining may be an instance not only of α but also of β ... Concepts in philosophy are specified (i.e. divided as genera into their constituent species) by an *a priori* distinction between possible "kinds" not unlike the *a priori* distinctions which specify the concepts of mathematics; but different from these in that the latter are *disjunctive*, the former *conjunctive*: the classes into which a non-philosophical concept can be sorted out are *exclusive*, those of a philosophical concept are *overlapping*. Thus all thought is *both* judgment *and* inference: all judgment *both* affirmative *and* negative ... This is very important for philosophical *method*. If anyone proposes to search for "a general characteristic which belongs in common to absolutely all right actions ... and which does not belong to any actions except those which are right" (Moore)[10] he is committed to all possible fallacies of philosophical method. (ibid.: 23–4)

Here Collingwood interpolates a sharp and rather acid criticism of Moore and those who follow his method:

> It is really an old doctrine, 'Quodlibet ens est unum, verum, bonum' is the same thing. Hence the sheer illiteracy of Moore's further question (as well as its ethical obtuseness) 'can we discover any characteristic which belongs in common to absolutely all good things and which does not belong to anything except what is a good?' (Moore 9). Yet Prichard, Ross, Carritt, habitually use this vicious method. (ibid.: 23–4)

In the same year, 1932, Collingwood wrote a new series of lectures on moral philosophy containing his most extensive methodological introduction yet. The following year the introduction was dropped as now being superseded by the publication of *An Essay on Philosophical Method*. As it does not add to what we have in *An Essay on Philosophical Method*, there is no need to dwell on it here—but it is worth observing that it contends that

> Philosophical thinking is *a priori*, and in this way is more like geometry than zoology. Therefore the specifications of its concepts must proceed, like that of mathematical concepts, by following the necessary implications of the concept itself, not by sorting the instances of it. But the resulting determinations will be mutually inclusive, not mutually exclusive: therefore instead of beginning with a concept and asking, what are the alternative ways in which this concept may be realised?—we must begin with a concept and ask, what are the necessary consequences entailed by the realisation of this concept? Each separate consequence will be a specification of the concept. (Collingwood 1932b: 8)

4 An Essay on Philosophical Method

Collingwood's views of method received their most complete and fully developed treatment in *An Essay on Philosophical Method*. We have seen something of how *An Essay on Philosophical Method* originated in the methodological introduction to his lectures on moral philosophy, but now we need to consider its content and argument. Let us begin with overlap: in philosophy the specific instances of concepts tend to overlap so that two or more concepts may be exemplified in the same instance: "any distinction in philosophy may be a distinction without a difference, (that is), where two philosophical concepts are distinguished Aristotle's formula may hold good, that the two are the same thing but their being is different" (EPM: 50). Aristotle's formula explains the essential insight lying behind the doctrine of the overlap of classes. The idea of a scale of forms develops this by suggesting that there is overlap not only in extension but also in intension. The subject matter of philosophy, owing to this overlap of classes, does not admit of classification into mutually exclusive species of a common genus of the sort found in the natural sciences. Philosophical concepts are generic; the species of a philosophical genus differ from each other both in degree and in kind; and in a philosophical scale of forms "the variable is identical with the generic essence itself" (ibid.: 60). Differences of degree between philosophical concepts cannot be measured because they are different both in degree and in kind (ibid.: 70–1); further, "if in philosophical thought every difference of kind is also a difference of degree, the specifications of a philosophical concept are bound to form a scale; and in this scale their common essence is bound to be realized differentially in degree as well as differentially in kind" (ibid.: 77).

The terms on a scale of forms are related by both opposition and distinction. Each term is distinct from its neighbours, but also opposed to it. This opposition is not absolute: "if the variable is identical with the generic essence, the zero end forms no part of the scale; for in it the generic essence is altogether absent. The lower end of the scale, therefore, lies not at zero, but at unity, or the minimum realization of the generic essence" (ibid.: 81). This might seem to imply that all opposition disappears and that we are left only with distincts, but in a philosophical scale of forms there is a fusion of distinction and opposition and therefore the scale does not consist merely of distincts (ibid.: 81–6). But opposition within a scale of forms does not imply the real existence of either end of the scale: "the lowest member of the scale, the minimum realization of the generic essence, is already, so far as it goes, a realization of this essence, and therefore distinct from other realizations; but, as the limiting case, it is an extreme, and therefore an opposite relatively to the rest of the scale" (ibid.: 82). Opposition appears at any point in the scale: "the same relation which subsists between the lowest member of the scale and the next above it reappears between any two adjacent forms. Each is good in itself, but bad in relation to the one above, and hence, wherever we stand on the scale, we are at a minimum point in it; and conversely, however far down we go, there is always the possibility of going lower without reaching absolute zero" (ibid.: 84). And infinity has no place in the scale: "each term in the scale … sums up the whole scale to that point. Wherever we stand in the scale, we stand at a culmination. Infinity as well as zero can thus be struck out of the scale, not because we never reach a real embodiment of the generic concept, but because the specific form at which we stand is the generic concept itself, so far as our thought yet conceives it" (ibid.: 89).

The classes of a philosophical concept overlap and "the higher term possesses not only that kind of goodness which belongs to it in its own right, but also the kind which originally or in itself belonged to its neighbour" (ibid.: 86–7). Each term has a double relation to its neighbours: "in comparison with the one below, it is what that professes to be, in comparison with the one above, it professes to be what it is" (ibid.: 87). This leads to the idea of determinate negation: "the higher term is a species of the same genus as the lower, but it differs in degree as a more adequate embodiment of the generic essence, as well as in kind as a specifically different embodiment" (ibid.: 88). The higher term is therefore not only distinct from the lower but also opposed to the lower. It "possesses not only its own specific character, but also that which its rival falsely claimed" (ibid.: 88). Negation is determinate because it both cancels and reaffirms: "the higher term thus negates the lower, and at the same time reaffirms it: negates it as a false embodiment of the generic essence, and reaffirms its content, that specific form of the essence, as part and parcel of itself" (ibid.: 88).

Finally, Collingwood further develops the notion of overlap and the scale of forms by explaining that the higher of any two adjacent forms overlaps the lower because it includes the positive content of the lower as a constituent element within itself, but it rejects the negative element in the lower, and this negative element is the denial that the generic essence contains anything more than the lower itself provides. It is this denial that constitutes its falsehood. "The lower overlaps the higher in a different sense: it does not include the higher as part of itself; it adopts part of the positive content of the higher while rejecting another part" (ibid.: 90). Taken together, "the overlap consists in this, that the lower is contained in the higher, the higher transcending the lower and adding to it something new, whereas the lower partially coincides with the higher, but differs from it in rejecting this increment" (ibid.: 91).

In *An Essay on Philosophical Method* Collingwood demonstrated to his own satisfaction that the logic of a philosophical scale of forms was possible. He had sought to do this since his rejection of realism. This demonstration hinged on the distinction between empirical or non-philosophical concepts as class concepts and philosophical concepts which (considered in their relations) displayed identity and difference, fusion of opposition and distinction, and fusion of degree and kind. Thus in a philosophical scale of forms the higher terms make explicit what is merely implicit in the lower. The starting point for Collingwood's understanding of method is Socratic. In philosophy we come "to know better something which in some sense we knew already" (ibid.: 11; cf. 161). This principle is re-asserted throughout the book, and is not per se distinctive (and Collingwood did not claim that it was). Collingwood's originality lies, rather, in his systematic elucidation of a conception of philosophical method in which this claim acts as a clue to the behaviour of philosophical concepts, which overlap in their instances; this, together with the principle that philosophy renders the implicit explicit, leads to the logic of the scale of forms.

Collingwood's achievement lay therefore in his attempt to think through clearly the requirements of a scale of forms which did justice both to his use of it and to his predecessors. It was the culmination of an engagement with Hegel, Plato, Bradley, Croce, and Gentile on the nature of a dialectical scale and the philosophical concept. To take an example: Collingwood did not mention Croce, but his presence can be felt because one of his concerns was to take account of Croce's criticism of Hegel's notion of dialectic. For Croce, because philosophical (universal) concepts are related by opposition, they exhibit dialectical relations; non-philosophical (individual) concepts are related only by distinction, hence they are non-dialectical. Collingwood's view is that this distinction, by which Croce discards much of Hegel's dialectic, overthrows the possibility of philosophy itself. In his view, it is because philosophical concepts are related by both opposition and distinction, and because they are both universal and categorical, that they are related dialectically and arrange themselves as a scale of forms. His conception of dialectic is a modification of both Croce's and Hegel's.

Collingwood reached his goal with a formulation of method which allowed him to accept the idea of a dialectical scale of degrees of knowledge. This enabled him to re-appraise the work of his predecessors, in particular, Bradley, because he was now able to admit Bradley's doctrine of degrees of truth and reality as valid—if understood as constituting a scale of forms. Previously Collingwood had rejected the doctrine of degrees of truth and reality, but now he was able to embrace them, re-interpreted through his conception of the scale of forms. In *Truth and Contradiction* he had submitted the theory of truth as coherence in the face of searching criticism. His main objection to the associated doctrine of degrees of truth and reality was that truth could not be arranged on a linear scale, a point repeated in *Speculum Mentis*, where he states that "the terms of a dialectical series are not related to one another in terms of degree, but by the assertion in each term of something which in the previous term was wrongly denied" (SM: 208). Following *An Essay on Philosophical Method* he was now free to use the notions of degrees of knowledge, truth, or reality principally because he had reconceptualized what is meant by the term 'degrees' in philosophy. This is made clear in *The Principles of Art*, where he remarked that Kant, in analysing imagination, approached the problem not by "trying to conceive real sensa and imaginary sensa as two co-ordinate species of the same genus", but "conceived the difference between them as a difference of degree" (PA: 187). He notes that "Here and elsewhere I use this word in the traditional philosophical sense, where differences of degree are understood as involving differences of kind; as in Locke's 'three degrees of knowledge', where each 'degree' is at once a fuller realization of the essence of knowledge than the one below (more certain, less liable to error) and also a fresh kind of knowledge" (ibid.: 187n).

A curious feature of *An Essay on Philosophical Method* is Collingwood's heavily qualified resurrection of the ontological argument in which he is not concerned to prove the existence of God as the object of religious worship but rather to make a claim about the peculiarity of philosophy as self-referential and hence committed to the existence of the object of philosophical thought. Because philosophical thinking is 'criteriological',[11] that is, it judges acts of thinking according to standards which in turn are self-reflexively applied to itself and because it is self-referential, being an instance of its subject matter, in its own performance it must exhibit the very qualities it is concerned to defend or criticize. The philosopher is therefore obliged to judge his or her own performance as a thinker. How is this related to the ontological argument? Collingwood claims that the argument proves that

> essence involves existence, not always, but in one special case, the case of God in the metaphysical sense: the *Deus sive natura* of Spinoza, the Good of Plato, the Being of Aristotle: the object of metaphysical thought. But this means the object of philosophical thought in general; for metaphysics, even if it is regarded as only one among the philosophical sciences, is not unique in its objective reference or in its logical structure; all philosophical thought is of the same kind, and every philosophical science partakes of the nature of metaphysics, which is not a separate philosophical science but a special study of the existential aspect of that same subject-matter whose aspect as truth is studied by logic and its aspect as goodness by ethics. Reflection on the history of the Ontological Proof thus offers us a view of philosophy as a form of thought in which essence and existence, however clearly distinguished, are conceived as inseparable. On this view, unlike mathematics or empirical science, philosophy stands committed to maintaining that its subject-matter is no mere hypothesis, but something actually existing. (EPM: 127)

This is rooted in the self-referentiality of propositions in logic: thus, in Ryle's paraphrase of Collingwood, logicians "enunciate principles of logic in propositions which themselves exemplify those principles. So their propositions exist. So the essence of the principles of logic involves the existence of examples of them" (Ryle 1935: 110). Further, Collingwood maintains that logic is neither purely normative nor purely descriptive, because

> A purely normative science would expound a norm or ideal of what its subject-matter ought to be, but would commit itself to no assertion that this ideal was anywhere realised. ... in logic the subject-matter is propositions, and the body of the science consists of propositions about propositions. ... the propositions of which logic consists must conform to the rules which logic lays down, so that logic is actually about itself; not about itself exclusively, but at least incidentally about itself. It follows that logic cannot be in substance merely hypothetical. ... logic, by existing, ... constitutes an actually existing subject-matter to itself ... Consequently no such discussion can be indifferent to the existence of its own subject-matter; in other words, the propositions which constitute the body of logic cannot ever be in substance hypothetical. A logician who lays it down that all universal propositions are merely hypothetical is showing a true insight into the nature of science, but he is undermining the very possibility of logic; for his assertion cannot be true consistently with the fact of his maintaining it. Similarly with inference. Logic not only discusses, it also contains reasoning ... logic has to provide not only a theory of its subject-matter, but in the same breath, a theory of itself; it is an essential part of its proper task that it should consider not only how other kinds of thought proceed, and on what principles, but how and on what principles logic proceeds. If it had only to consider other kinds of thought, it could afford to deal with its subject-matter in a way either merely normative or merely descriptive; but towards itself it can only stand in an attitude that is both at once. It is obliged to produce, as constituent parts of itself, actual instances of thought which realise its own ideal of what thought should be. Logic, therefore, stands committed to the principle of the Ontological Proof. Its subject-matter, namely thought, affords an instance of something which cannot be conceived except as actual, something whose essence involves existence. (EPM: 129)[12]

5 An Essay on Philosophical Method and After

After the publication of *An Essay on Philosophical Method*, Collingwood employed or illustrated his use of the scale of forms. It is to be found in his 'Notes Towards a Metaphysic', and in lectures such as 'Method and Metaphysics'; it is employed historically in *The Idea of Nature* and *The Idea of History*; it is present in *The Principles of Art* and forms the spine of *The New Leviathan*.

In *An Essay on Philosophical Method* Collingwood does not distinguish or systematically explore all of the possible applications of the scale of forms. For example, in his work we find consciousness, forms of experience, concepts, and historical development all variously arranged as scales of forms. But this prompts the question: does scale of forms analysis require modification if applied to these different objects? *An Essay on Philosophical Method* is silent on this, perhaps because it implicitly assumes that there are no essential differences. Although it could be argued that use of the method in different contexts requires modification, it should be noted that neither in *An Essay on Philosophical Method* nor later did Collingwood indicate that he saw any need for fundamental modification to its argument.

The Principles of Art was quietly unobtrusive in its employment of the scale of forms and the concept of overlap. We have already noted one passage clarifying the notion of degrees in philosophy; another is where Collingwood refers to the overlap between art and craft and urges readers to avoid the fallacy of precarious margins (PA: 22n). This point is often missed by hostile critics who frequently take Collingwood to be claiming that what is craft is *ipso facto* not art and vice versa. As we saw earlier, in his criticism of Moore's ('and his followers') method in moral philosophy, this is the precise opposite of what he intended. In 'What Civilization Means' (1939–40) the method is employed in an explicit rebuttal of historical relativism. Collingwood analysed the ideals of civilized conduct and demonstrated that their inter-relations are those of members in a scale of forms. There cannot be, therefore, a single unilinear scale in which civilizations differ only in degree: civilizations differ both in degree and in kind. Generally, though, Collingwood tended not to draw attention to method, preferring to let the dialectic emerge as the argument unfolded. There remains a curiosity, however: for the scale of forms was neither obviously employed nor mentioned in *An Essay on Metaphysics*, where, one might expect it to have been. We examine this curiosity below. The last book Collingwood published in his lifetime was *The New Leviathan* which drew extensively on themes from his lectures on moral philosophy and is explicitly arranged as a scale of forms. However, it makes an apparent modification to the doctrine by introducing 'the law of primitive survivals', stating that "when A is modified into B there survives in any example of B, side by side with the function B which is the modified form of A, an element of A in its primitive or unmodified state" (NL: 9.51).[13]

T.M. Knox reports that, after completing the *Essay*, Collingwood remarked that having "propounded a theory of philosophical method, he was now proceeding to apply it to a problem which had never been solved, namely, to the philosophy of nature" (IN: v). To this end he began a series of notebooks—'Notes Towards a Metaphysic'. In his opening remarks Collingwood made the relationship between them and *An Essay on Philosophical Method* clear:

> The main principle of my method is the Scale of Forms—in this problem we have a genus (I suppose its name is Reality) divided into a scale of species Matter, Life, Mind. According to the method, these should be *kinds* of reality and also *degrees* of reality, the highest the most real: each should be *distinct* from the next and also *opposed* to the next ... each should sum up the whole scale to that point and from its own point of view be identical with the genus, so that, from the point of view of a lower term (e.g. Matter) that term *is* the genus (Reality) and the higher terms (Life, Mind) are simply nothing—the names of errors. Whereas from the point of view of a higher term (e.g. Mind) although that term *is* the genus (Mind=Reality) the same term includes the lower terms (Matter, Life) and, instead of denying these, asserts them as implicates of itself. It does deny them in one sense, but only in the sense that it denies their denials, viz. Their claim to be the only or highest term in the scale. (Collingwood: 1933–4: 1)

This goes beyond *An Essay on Philosophical Method in asserting* that reality constitutes a scale of forms. But is Collingwood asserting that reality *in re* constitutes a scale of forms or that reality, *qua* philosophical concept, necessarily articulates itself as a scale of forms? *An Essay on Philosophical Method* had confined itself to the analysis of concepts and remained agnostic on these wider issues. Further, he introduces a new metaphysical principle by asking what drives movement from one point on the scale to another: "what about the *nisus*? i.e. what is the nature of the force which drives us from a lower term to the next higher, or if you like converts that into this? ... It is what Spinoza would call an immanent causality: something in the first term which converts it into the second" (Collingwood 1933–4: 1).

In 1935 Collingwood, in a chapter 'Method and Metaphysics', showed how scale of forms analysis could be applied to a metaphysical question, the general nature of reality. He endeavoured to show that reality is not a class concept but a philosophical concept to which the rules of classification appropriate to the empirical or the exact sciences do not apply.

The possibility of metaphysics was thereby defended on the grounds that (because reality is not a class concept) metaphysics as the search for the general nature of reality is not an empty search for a bare abstraction. Collingwood explains that

> I should expect the concept of reality, like other philosophical concepts, to exhibit the structure of a scale of forms, where the different kinds of real things differ both in the kind and in the degree of reality that they possess; a scale nowhere going down to zero, so that there is no such thing as the completely unreal, but one in which every term is relatively unreal by contrast with higher terms, except the highest of all, and I should not expect that we could know, except in a very dim and imperfect way, what the absolutely highest term is. (Collingwood 1935: 8)

This seems to imply that there is no distinction in philosophy between the concept of reality and reality itself. Our understanding of reality is necessarily conceptual and therefore reality, as we understand it, is arranged as a scale of forms. Collingwood's idealism, revealed here, was recessive (although implicit) in *An Essay on Philosophical Method* but emerges more strongly in some of the manuscripts following publication.

6 A Tale of Two Essays

An Essay on Metaphysics neither refers to *An Essay on Philosophical Method* nor (on the surface, at least) employs its method. The significance of this lies in the fact that some commentators have argued that between the writing of the two Collingwood fundamentally changed his mind on the relation between philosophy and history and embraced a historical relativism in which history swallowed philosophy whole. Clearly Collingwood's thought underwent change and development over time, but was there a radical change of the sort alleged?

First, Collingwood never referred directly or obliquely to such a change, and all of his later references to *An Essay on Philosophical Method* are positive. It is nonetheless remarkable that *An Essay on Metaphysics* contains no reference to *An Essay on Philosophical Method*. This is all the more remarkable because Collingwood insisted that the two *Essays* should be regarded as companions. Of course, Collingwood could have been radically mistaken about the import of his own doctrines and failed to recognize the change in attitude. But this is surely implausible. Again, there is no continuity of vocabulary between the two *Essays*, but this is typical rather than exceptional: throughout his philosophical career he fashioned his thinking and its expression anew for each philosophical occasion, problem, and audience.

Secondly, absolute presuppositions (which the metaphysician is allotted the task of discovering and cataloguing) clearly have a close affinity with what Collingwood earlier termed philosophical concepts or transcendentals. They are non-empirical concepts concerning the general nature of reality; they are both universal (not aggregative or enumerative) and categorical. They are not generalizations derived from experience but yardsticks brought to bear on experience.

Thirdly, it is worth considering another level of connection between the two *Essays*. Both were regarded by Collingwood as examinations of the presuppositions of philosophy itself, and he regarded such an examination as vital to philosophy as a self-reflective discipline. In *An Essay on Philosophical Method* this emerges most strongly in his criticism of the failure of the analytical school to address the issue of their own presuppositions. There he argues that the methods and procedures of the analytic school rest on unacknowledged and unanalysed presuppositions which need to be brought to light. Just as *An Essay on Philosophical Method* is a self-reflective account of philosophy directed upon itself, an examination of the presuppositions of philosophy and its method, so *An Essay on Metaphysics* is also philosophy of philosophy. Like *An Essay on Philosophical Method* it is consciously self-referential; in some ways it operates at a deeper level of analysis: in *An Essay on Philosophical Method* Collingwood *assumed* the existence of philosophical concepts; in the later *Essay* he sought to prove the existence of this class of propositions by demonstrating that absolute presuppositions can neither be regarded as analytically true nor derived empirically.

One of Collingwood's central points in both *Essays* is that reliance on presuppositions is inescapable and that this also applies to philosophy itself: hence his critical comments in *An Essay on Philosophical Method* on the failure of the emerging analytical school to provide an account of their own presuppositions. Further, he maintained, as applied to philosophy they give rise to the peculiarities of self-instantiation and self-reference. This issue is later also addressed directly in *An Essay on Metaphysics*: "because all science begins with a question (for a question is logically prior to its own answer) all science begins with a presupposition. Metaphysics, therefore, either has presuppositions or is no science. The attempt at a metaphysics devoid of presuppositions can only result in a metaphysics that is no science, a tangle of confused thoughts whose confusion is taken for a merit" (EM: 63). Metaphysics, therefore, needs to acknowledge the presuppositions on which it rests: it cannot rest content with examining the presuppositions of other forms of inquiry. Its own presuppositions concern the nature of metaphysics itself and hence are both subject and object in metaphysical thought. Elsewhere Collingwood elaborated this point:

> An inquiry into the nature of metaphysics is a metaphysical inquiry [this] consists of an historical (critical) examination of the presuppositions of metaphysicians, with particular regard to their presuppositions as to the nature of metaphysics itself. The inquiry is an effort to arrive at a definition of metaphysics. Yet this definition when finally stated will be found to be one of the presuppositions of the very inquiry itself, a presupposition which has determined the method of procedure ... In metaphysics, the method of thought becomes the object of thought, i.e. object and method of study become unified. There is no external, independent subject. (Collingwood 1939)

7 Relativism and the Radical Conversion Hypothesis

Lionel Rubinoff was the originator of the term 'radical conversion hypothesis'. He sought to show that Collingwood's thought developed through the elaboration of themes already present in the early works, principally *Speculum Mentis*. He argues that Collingwood's thought is "not characterised by a radical conversion from 'idealism' to 'historicism'" but that the rapprochement he set out to accomplish in 1916 was achieved in *Speculum Mentis* and *An Essay on Philosophical Method* and subsequently applied to the interpretation of philosophical problems in his later writings (Rubinoff 1970: 24). A fellow traveller in rejecting the radical conversion hypothesis was Louis Mink (1969). Both agree in rejecting the claim that Collingwood became a historical relativist, denying the possibility of transhistorical truth and asserting that each epoch of human thought was bounded by a set of absolute presuppositions beyond the reach of rational criticism.

Despite this, T.M. Knox, Collingwood's student and friend, was not lacking apparent evidence in making the claim that there was a fundamental change in Collingwood's thought. He grouped his writings into three: (1) *Religion and Philosophy* and *Speculum Mentis*; (2) *An Essay on*

Philosophical Method, *The Idea of Nature* and the bulk of *The Idea of History*. (These were, in Knox's estimation, the best, typified by a balanced view in which philosophy and history were understood as related but distinct categories); (3) *An Autobiography*, *An Essay on Metaphysics* and *The New Leviathan* all marred by historicism and scepticism. *The Principles of Art* was akin in part to the second group, in part to the third. For Knox, Collingwood betrayed philosophy by absorbing it into history, and he found himself "compelled to believe that his philosophical standpoint radically changed between 1936 and 1938" (IH: xi).

Let us consider some of the evidence cited by Knox. What, for example, did Collingwood mean when he wrote that "Philosophy as a separate discipline is liquidated by being converted into history"? (IH: x). The crucial word is 'separate'. It is philosophy as a *separate* discipline which is rejected: and what we have is an overlap between philosophy and history. The attempt to maintain philosophy and history as separate disciplines commits the 'fallacy of precarious margins': this lies in assuming that "the overlap which has already affected a certain area of the class in question can be trusted not to spread, and that beyond its limit there lies a marginal region in which the instances exhibit only one of the specific forms, uncontaminated by the presence of the other" (EPM: 48). But did Collingwood commit the opposite fallacy of 'identified coincidents' (ibid.: 49), which is the view that because two concepts overlap therefore they are identical? Elsewhere he writes of the science of mind as a special kind of history: "The absorption is mutual: the product is not philosophy based on history nor history based on philosophy, it is both these things at once" (Collingwood 1937a). This implies that far from identifying history and philosophy Collingwood took philosophy and history to be inseparable and sought to avoid both of the fallacies he identified in *An Essay on Philosophical Method*.

Additional evidence cited by Knox includes Collingwood's remarking in a letter to him that "St. Augustine looked at Roman history from the point of view of an early Christian; Tillemont, from that of a seventeenth-century Frenchman; Gibbon, from that of an eighteenth-century Englishman; Mommsen, from that of a nineteenth-century German. There is no point in asking which was the right point of view. Each was the only one possible" (IH: xii). This is taken by Knox to be unmitigated historical relativism. But is it? Partly it depends on the interpretation of 'point of view'. If it is taken to mean that the conclusions of historians are determined in every detail by their epoch and situation, the charge would stand;

however, if it means that every historian works within an inherited framework of problems and solutions, it does not. It is the difference between what Stephen Toulmin calls relativism and relationism. Further, there is nothing in this passage that Collingwood does not assert in his 1930 pamphlet, *The Philosophy of History*, written prior to the period in which Knox asserts that the radical change of mind occurred. There he states that historians inescapably possess a point of view, and the jibe that "the attempt to eliminate this 'subjective element' from history is always insincere—it means keeping your own point of view while asking other people to give up theirs—and always unsuccessful. If it succeeded, history itself would vanish" (Collingwood 1930: 15). Either one claims that Collingwood was a historicist in 1930 (which Knox does not maintain) or one has to accept that there was no conversion.

In his modified advocacy of the radical conversion hypothesis, Donagan agrees with Knox that the *Autobiography* and *An Essay on Metaphysics* marked a shift to historicism, but revises his account. He agrees that Collingwood underwent a 'radical conversion' between 1936 and 1938, but he disagrees with Knox on both the explanation and the division of the later writings into the 'non-historicist' and the 'historicist' (Donagan, 1962: 12ff).

However, although Collingwood underwent no radical conversion, this is not to deny that he felt the pull of relativism or relationism. How could it be otherwise for a sensitive historian with an interest in anthropology? Appreciating relativity may seem only a step away from adopting relativism, but the step is not necessarily taken. Collingwood was alive to this and sought to do justice to both the unchanging and changing elements in knowledge and experience. To be satisfied with differences in degree alone implies a single unilinear scale; to be satisfied with sheer difference is to abandon the search for unifying principles and would amount to relativism. Collingwood developed, as we have seen, a conception of method which accommodated differences both in degree and in kind, thereby avoiding both absolutism and relativism.

Thus, the opponent of the 'radical conversion hypothesis' can readily admit that Collingwood was exercised by the relativity of thought and action in history, archaeology, and anthropology. How far this went can be seen in his comments that: "We all know what kind of world it is to which you refer as the reality (or objective reality) world. But what exactly is its *title* to the name reality? I am haunted by a suspicion that it has none except the fact that adult members of modern European society (or rather, the 'educated classes' of that society) are agreed to treat it as real; that (in other words) it is the *conventional* world of a particular historically-determined culture. It is *not* the world of adult human beings *as such*" (Collingwood 1937b). This appears to be teetering on the edge of relativism. The passage was written when Collingwood had been working extensively on anthropology and folklore and on a book-length manuscript on anthropology. It is also clear that a strong influence was exerted by Evans-Pritchard's *Witchcraft, Oracles and Magic Among the Azande*, which he read in manuscript and immediately quoted in print (PA: 8n).

However, Collingwood resisted the temptation, and in the middle of the disputed period he wrote a clear refutation of historical relativism. In 'What Civilisation Means' he steers a course between 'historical monism' and 'historical pluralism' and argues that

> the present-day conception of the relation between civilization and barbarism is that any given society at any given time has its own standard of civilized life, and thinks of itself as civilized in so far as it recognizes that standard, and of other societies as barbarous in so far as they do not. The attempt to be a civilized European is an attempt to live one kind of life; the attempt to be a civilized Chinaman is an attempt to realize a life of another kind. It is not surprising that Europeans think Chinamen barbarous and that Chinamen think the same of Europeans. Each rightly thinks that the other when judged by his standard, fails to satisfy it. The question with regard to any given society, then, is not how high it stands in the scale of civilization, for there is no one scale; still less whether it is just civilized or just barbarous, for every society is civilized, or it would not be a society: but in what way it is civilized. And, from the point of view of any one civilization, any other is merely one of the innumerable forms of barbarism. (NL: 489)

Collingwood acknowledges that this is, prima facie, relativism:

> This may seem a dangerous option. It may seem to imply that for any given society the proposition 'we are civilized' has a sense peculiar to that society, for 'civilized' has no constant meaning in such propositions; the definition of civilized conduct varies from society to society, and in any given society from time to time, in such a way that every society thinks its own present way of behaving civilized and the others barbarous. This is called 'historical relativism', and is rightly regarded with suspicion, because it really amounts to denying what it professes to explain. It amounts to denying that there is any such thing as an ideal of civilized conduct: not merely that there is one single ideal valid for all societies and all times, but that there are many ideals each valid for one society at one time. (NL: 489)

He then provides a solution rooted in the conception that civilizations differ in both degree and kind and cannot be placed on a scale differing merely in degree. Collingwood, although tempted by what was plausible in relativism, saw its dangers and avoided them by employing what, in old idealist language, might have been termed identity in difference or the logic of the concrete universal. His language was different, but the aim was the same. How did he attempt this? He argues that each civilization lives up to its own ideal in both degree and in kind, but fails to live up to it perfectly and that this partial success reveals the existence of another order of ideal:

> Every man who is civilized in a certain way and up to a certain degree recognizes other ways and other higher degrees in which he might be civilized, and is to that extent conscious of shortcomings in his own civilization. This implies that, in addition to the ideal of civilization which he both recognizes and realizes, he has another ideal of civilization which he recognizes but does not realize. Let us call this an ideal of the second order. (NL: 491–2)

Ideals of the first order are ideals which are also social facts; they are both recognized and realized in the conduct of people and societies. By first-order ideals Collingwood means our normal everyday customs and manners such as the way in which we express politeness. But if there are standards or criteria by which we judge our thought and actions, this implies the possibility that our thoughts and deeds may not realize the ideal which it sets itself. Here we recognize an ideal of conduct which we do not realize in our actions and this generates ideals of the second order.

> How widely this second order of ideals differs from the first will … vary enormously from man to man and from society to society; and in the same man or society from time to time … every man who is conscious of realizing one ideal of civilized life is also conscious of entertaining a second ideal which he does not realize. Ideals of the first order, which are also social facts, differ among themselves … indefinitely. (NL: 492)

For Collingwood there are three ways of conceiving of differences in civilization: the first in which we divide civilized from uncivilized absolutely; the second in which we treat them as constituting a scale on which each differs from its neighbour merely by degree; and the third in which they constitute a scale in which each differs from its neighbours both in degree and in kind. Collingwood suggests that we can fit these together by using the distinction between different orders of ideals: "The first implied a hard and fast line between being civilized and being uncivilized", for example "when 'civilized' means 'civilized' according to the ideal not only recognized but realized in a certain social group", such as taking your hat off when entering a house. The second approach implies a difference of degree and: "One action is called civilized and another barbarous, by one and the same ideal of civilized conduct, in so far as they realize this ideal in a higher or a lower degree". The third implies a difference of kind, for example between European civilization or Chinese civilization. Collingwood then draws together degree and kind in his analysis of three orders of ideal:

> Ideals of the first order, we have seen, may vary indefinitely. Ideals of the second order may also vary indefinitely. But they may vary less than ideals of the first order. It is a fact that the line between civilized and barbarous conduct is drawn in different places by different groups of English people; but there is a tradition to the effect that many of these groups recognize, even if they do not all realize, a single ideal which is both recognized and realized by one group: the ideal of conduct as 'gentlemanly'. How far this tradition is based on fact I do not ask; I only point out that the people who think it is believe that the idea of a 'gentleman' is a first order ideal for one group of Englishmen and a second order ideal for other Englishmen or even for all Englishmen. Similarly, it used at one time to be thought … that the French ideal of civility is an ideal valid for all Europeans. This implies that French civility is a first order ideal for all Frenchmen, and a second order ideal for others … But even second order ideals are only partial. (NL: 492–3)

Collingwood now considers what there is about first- and second-order ideals which justify us in calling them ideals of the same thing. For him different ideals of civilized conduct, although divergent, must also be in some sense convergent, for they are both ideals of the same thing, and, he suggests, "unless this phrase means that in some way they are the same ideal it means nothing. Thus the historical pluralism of the present day does not exclude a sense in which all civilization is one" (NL: 489). The distinction between different order of ideals is an attempt to embellish this statement. Collingwood, after having distinguished first- and second-order ideals of civilized conduct distinguishes a third-order ideal which is a truly universal and universalizable ideal, and the logical ground of first- and second-order ideals. He characterizes this third-order ideal as follows:

> There is an ideal of civilized conduct in which … elements of particularity are left behind: an ideal of universal civility; civility on every kind of occasion, civility under any kind of provocation, civility to every kind of person … This … third order ideal, is logically the source of all other ideals of civilized conduct. Not that it was reached before them; on the contrary, civility began, no doubt, at home; but on reflection it became evident that being civil to certain kinds of person, on certain kinds of occasion, and under certain kinds of provocation, is only a special case or … a 'mode' of being civil: and that the particular ideals of civility which are realized, or recognized without being realized, by this or that particular logically presupposes the ideal of civility as such. This is the sense … in which all civilizations, or ways of living in a civilized manner, are one. They are one in the sense that they all converge upon this one ideal. It must be one ideal, there can be no other beyond it, because it is absolutely unqualified. There can be no fourth order ideal of civilized conduct between man and man. Every man must recognize it when he reflects on his own life as a member of a civilized society. He finds himself behaving civilly in certain limited ways. On reflection, he discovers that he is trying to behave civilly in other and more far-reaching, but still limited, ways. On further reflection, he discovers that he is trying to behave civilly without any limit whatever. (NL: 493–4)

Collingwood asks: "Why is the second order ideal not realized?" The answer he gives indicates the relation between these three order of ideals and the scale of forms; and they also show how scale of forms analysis can be used to allow, on the one hand, for the relativity of things (their difference in kind), while on the other hand insisting on their essential unity, and thus avoiding the pitfalls of relativism.

Why, then, is the second-order ideal not realized? Because "to recognize an ideal is to try to realize it: recognizing it is not just an intellectual act but an act of will"; the attempt to realize the second order does not entirely fail: "The second order ideal, considered simply as an ideal, includes the first order ideal as part of itself; realization of the first order ideal, therefore, is a partial realization of the second order ideal. The failure to realize the second order ideal is failure only to realize that part of the second order ideal wherein it goes beyond the first order ideal" (NL: 495). A second-order ideal, once realized, would become a first-order ideal: it would become a social fact. Again, a first-order ideal, if it ceased to be fully realized would become a second-order ideal: no longer a social fact, but an ideal recognized but not realized. On Collingwood's account, first- and second-order ideals are what they are only relatively and not absolutely. Furthermore, they overlap: the second-order ideal, as an ideal, contains the first-order ideal as a part of itself; realization of the first-order ideal is, therefore, at the same time a partial realization of the second-order ideal. When we say that we fail to realize a second-order ideal we mean not that we do not realize it *at all*, but that we only realize it to a limited extent: where we fail is in not realizing that part of the second-order ideal in which it goes beyond the first-order ideal.

This account can be developed through consideration of the formulation for the overlap of classes as expressed in *An Essay on Philosophical Method* which states that

> The higher of any two adjacent forms overlaps the lower because it includes the positive content of the lower as a constituent element within itself. It only fails to include the lower in its entirety because there is also a negative aspect of the lower, which is rejected by the higher: the lower, in addition to asserting its own content, denies that the generic essence contains anything more, and this denial constitutes its falsehood ... The lower overlaps the higher in a different sense: it does not include the higher as part of itself, it adopts part of the positive content of the higher, while rejecting another part. (EPM: 90)

Comparison of this with the relation between first- and second-order ideals shows that the second-order ideal overlaps the first-order ideal in precisely the same way as the higher of two adjacent terms overlaps the lower and vice versa. The relation between the two is the same as that between adjacent members on a scale of forms. But, if first- and second-order ideals are adjacent members on a scale of forms, what is a third-order ideal? The answer is that the third-order ideal is not a still higher term on a scale of forms, but the philosophical concept which is the logical ground of the scale itself: the genus of which each term in the scale is a specification. That is, if our concept is civilization as an ideal, this concept will articulate itself as a scale of forms in which first- and second-order ideals will be alternative names for any two adjacent members of the scale, and in which the third-order ideal will be the concept which the specifications logically presuppose. First-order ideals may be fully realized so far as they go; second-order ideals are not fully realized; third-order ideals are in one sense fully realized and in another not: they are realized in so far as any ideal of civilization is realized, and not realized in so far as no realization of the ideal of civilization is complete or perfect. In *An Essay on Philosophical Method* Collingwood asserts that in a philosophical scale of forms "the variable is identical with the generic essence itself" (EPM: 60) and that each term is, therefore, a specification of the generic essence, but as compared with its successor, a relatively inadequate one; and as compared with its predecessor a relatively adequate one. Each term in the scale sums up the scale to that point. Wherever we stand in the scale we stand at an end, but only at a relative end. This is because the specific form at which we stand is the generic concept itself, so far as we yet conceive it. What is meant by saying that such an ideal is presupposed by our first- and second-order ideals is not that we know the way in which such an ideal (universally realized) would empirically manifest itself but simply that the notion of such an ideal is implicit in our everyday conduct and descriptions of different ideals of civilized conduct as ideals of one and the same thing—civility itself.

8 Conclusion

The primary focus of this chapter is neither controversy nor critique.[14] What it has sought to achieve is to place Collingwood's reflections on philosophical method at the centre of his philosophical development and to show that he was always (at the very least) methodologically alert and (frequently) methodologically much more self-aware than his contemporaries, whether philosophical friends or foes. It has also addressed the issue of historical relativism and of the purported 'radical conversion' by showing that Collingwood remained true in his last writings to the conception of philosophical method he developed over many years culminating in its mature expression in *An Essay on Philosophical Method*.

Acknowledgments I would like to acknowledge the British Academy and Leverhulme Trust (Senior Research Fellowship SF160132) for their support in carrying out this research.

Notes

1. This chapter draws on material from Connelly and D'Oro (2005) and Connelly (2013).
2. For canonical statements of the mantra see Prichard, *Kant's Theory of Knowledge*, 115–19; Cook Wilson, *Statement and Inference*, 802.
3. See also Collingwood's use of Vico's principle in NL, 1.33 & chapter 9.
4. The two fallacies (which I have dubbed 'lumping' and 'splitting') were called by Collingwood the 'fallacy of identified coincidents' and 'the fallacy of precarious margins' (EPM: 47–8).

5. Not that he was always happy with this label—see, for example, his sharp riposte to Gilbert Ryle (EPM: 255–7).
6. For the relation of Collingwood's thought to that of the Italian neo-idealists, see Peters 2013.
7. For the fullest statement of Gentile's philosophy, see *The Theory of Mind as Pure Act*, 1922.
8. For details of the relationship between Collingwood and de Ruggiero, see Connelly 1995, 2020.
9. See *The New Leviathan*.
10. Moore 1912: 8–9.
11. He introduced the term in *The Principles of Art*.
12. These claims led both to a riposte by Gilbert Ryle in *Mind* (with a reply by E.E. Harris) and to extensive philosophical correspondence between Collingwood and Ryle. For a full account, see Connelly and D'Oro 2005.
13. See Connelly and D'Oro, xlviii.
14. For assessments and critique of Collingwood's view on philosophical method, see Martin 1974; Ducasse 1936 and Donagan 1972.

Bibliography

Carritt, E.F. 1928. *The Theory of Morals*. London: Oxford University Press.

Collingwood, R.G. 1916a. The Devil. In *Concerning Prayer*, ed. B.H. Streeter and L. Dougall. London: Macmillan.

———. 1916b. *Religion and Philosophy*. London: Macmillan.

———. 1917. *Truth and Contradiction*, Chapter Two, Bodleian Library, 16/1.

———. 1918. Notes on *Religion and Philosophy* and *The Devil*, in Private Possession.

———. 1920a. *Libellus de Generatione*, Bodleian Library, Dep Collingwood, 28.

———. 1920b. Notes on Hegel's Logic, Bodleian Library, Dep Collingwood, 16/2.

———. 1920c. Sketch of a Logic of Becoming, Bodleian Library, Dep Collingwood, 16/3.

———. 1920d. Draft of Opening Chapters of a "Prolegomena to Logic" (or the Like) Bodleian Library, Dep Collingwood, 16/5.

———. 1921. Croce's Philosophy of History. *Hibbert Journal* 19: 263–278.

———. 1922. *Ruskin's Philosophy*. Kendal: Titus Wilson & Son.

———. 1923a. Can the New Idealism Dispense with Mysticism? *Proceedings of the Aristotelian Society Supplementary Volume* 3: 161–175.

———. 1923b. Science "Greats". *Oxford Magazine*, March 15: 301–2.

———. 1923c. Science Greats. *Oxford Magazine*, May 10: 340.

———. 1923d. 'Action': Lectures on Moral Philosophy, Bodleian Library, Dep Collingwood, 3/1.

———. 1924. *Speculum Mentis*. Oxford: Clarendon Press.
———. 1925. Nature and Aims of a Philosophy of History. *Proceedings of the Aristotelian Society*, New Series, 25 (1924–5): 151–74.
———. 1926. Economics as a Philosophical Science. *International Journal of Ethics* 36 (2): 162–185.
———. 1927. The Idea of a Philosophy of Something, and, in Particular, a Philosophy of History, Bodleian Library, Dep Collingwood, 14/2.
———. (1929a). Political Action. *Proceedings of the Aristotelian Society*, New series. 29 (1928–9): 155–176.
———. 1929b. Lectures on Moral Philosophy, Bodleian Library, Dep Collingwood 10.
———. 1930. *The Philosophy of History*, Historical Association. London: Bell.
———. 1932a. Utility, Right and Duty, Bodleian Library, Dep Collingwood 6/3.
———. 1932b. Lectures on Moral Philosophy, Bodleian Library, Dep Collingwood, 7.
———. 1933–4. Notes Towards a Metaphysic, Bodleian Library, Dep Collingwood, 18/3.
———. 1937a. Letter to Guide de Ruggiero, 12 June 1937. Istito de Storia Moderna, Rome, Photocopies in the Bodleian Library, Oxford, Dep. Collingwood 26.
———. 1937b. Letter to Margaret Lowenfeld, 19 May 1937. Wellcome Library, Special Collections PP/LOW/I/6.
———. 1938. *The Principles of Art*. Oxford: Clarendon Press.
———. 1939. Notes on Collingwood's Lectures on Metaphysics by Louis Hector.
———. 1992. *The New Leviathan* [1942], Revised Edition with Additional Material. Ed. D. Boucher. Oxford: Oxford University Press.
———. 1993. *The Idea of History* [1946], Revised Edition with Additional Material. Ed. W.J. van der Dussen. Oxford: Oxford University Press.
———. 1998. *An Essay on Metaphysics* [1940], Revised Edition with Additional Material. Ed. R. Martin. Oxford: Oxford University Press.
———. 2005. *An Essay on Philosophical Method* [1935], Revised Edition with Additional Material. Ed. J. Connelly and G. D'Oro. Oxford: Oxford University Press.
———. 2013. *An Autobiography* [1939] *and Other Writings*, with Essays on Collingwood's Life and Work, Edited with an Introduction by D. Boucher and Teresa Smith. Oxford: Oxford University Press.
Connelly, J. 1995. Art Thou the Man? Croce, Gentile or De Ruggiero. In *Philosophy, History and Civilization: Essays on R.G. Collingwood*, ed. D. Boucher, J. Connelly, and T. Modood. Cardiff: University of Wales Press.
———. 2013. Collingwood Controversies. In *An Autobiography and Other Writings*, ed. R.G. Collingwood. Oxford: Oxford University Press.

———. 2020, Forthcoming. Guido de Ruggiero's Relationship with British Idealism. *Collingwood and British Idealism Studies* 26.

Connelly, J., and G. D'Oro. 2005. Introduction to Revised Edition of *An Essay on Philosophical Method*. Oxford: Oxford University Press.

Connelly, J., P. Johnson, and S. Leach. 2015. *R.G. Collingwood: A Research Companion*. London: Bloomsbury.

Cook Wilson, J. 1926. *Statement and Inference*. Oxford: Clarendon Press.

Croce, B. 1913. *Philosophy of Giambattista Vico*. Trans. R.G. Collingwood. London: Howard Latimer.

Donagan, A. 1962. *The Later Philosophy of R.G. Collingwood*. Oxford: Clarendon Press.

———. 1972. Collingwood and Philosophical Method. In *Critical Essays on the Philosophy of R.G. Collingwood*, ed. M. Krausz. Oxford: Clarendon Press.

Ducasse, C.J. 1936. Mr. Collingwood on Philosophical Method. *Journal of Philosophy* 33 (4): 95–106.

Gentile, G. 1922. *The Theory of Mind as Pure Act*. Trans. H.W. Carr. London: Macmillan.

Jones, H. 1917. Report on Collingwood's *Truth and Contradiction*, Macmillan Archives.

Joseph, H.W.B. 1931. *Some Problems in Ethics*. Oxford: Clarendon Press.

Martin, R. 1974. Collingwood's *Essay on Philosophical Method*. *Idealistic Studies* 4: 224–250.

Mink, L.O. 1969. *Mind, History and Dialectic: The Philosophy of R.G. Collingwood*. Bloomington: Indiana University Press.

Moore, G.E. 1903. *Principia Ethica*. Cambridge: Cambridge University Press.

———. 1912. *Ethics*. London: Williams and Norgate.

Muirhead, J.H. 1932. *Rule and End in Morals*. London: Oxford University Press.

Peters, R. 2013. *History as Thought and Action: The Philosophies of Croce, Gentile, de Ruggiero, and Collingwood*. Exeter: Imprint Academic.

Prichard, H.A. 1909. *Kant's Theory of Knowledge*. Oxford: Clarendon Press.

———. 1912. Does Moral Philosophy Rest on a Mistake? *Mind* 21 (81): 21–37.

———. 1928. *Duty and Interest*. Oxford: Clarendon Press.

———. 1932. *Duty and Ignorance of Fact*. Oxford: Clarendon Press.

Rubinoff, L. 1970. *Collingwood and the Reform of Metaphysics: A Study in the Philosophy of Mind*. Toronto: University of Toronto Press.

Ryle, Gilbert. 1935. Mr. Collingwood and the Ontological Argument. *Mind* 44: 137–151. Reprinted in *Collected Essays*, vol. 2. London: Hutchinson, 1971.

Stocks, J.L. 1932. *The Limits of Purpose*. London: Benn.

Toulmin, Stephen. 1972. *Human Understanding*. Vol. I. Princeton: Princeton University Press.

Vico, G. 1948. *The New Science*. Trans. T.G. Bergin and M.H. Fisch. New York: Cornell University Press.

CHAPTER 4

Collingwood on the Relationship Between Metaphysics and History

Jan van der Dussen

1 Introduction

Collingwood's *An Essay on Metaphysics* is, besides *The Idea of History*, perhaps his best-known philosophical book, but also a book that has aroused lively discussions. The nature of its reception shows a peculiar character, however, since the opinions on some of the most crucial aspects of the book widely diverge. For instance, in contrast to Collingwood's explicit position that metaphysics should be considered a historical science, it has been argued that in fact his theory of absolute presuppositions, as explained in the first four chapters of the book, is his central theme. Collingwood is partly to blame for the confusion raised about the nature of *An Essay on Metaphysics*. For his thesis that metaphysics is a historical science is not so much argued or substantiated in it, but rather taken for granted. Besides this, his views on the nature of the study of history are implied as well, they also being taken as a matter of course.

In January 1938 Collingwood wrote an essay entitled 'Function of Metaphysics in Civilization' in preparation for lectures, which is included in the revised edition of *An Essay on Metaphysics* (1998). It is a highly interesting document, especially since he develops in it an extensive

J. van der Dussen (✉)
The Open University, Milton Keynes, UK

K. Dharamsi et al. (eds.), *Collingwood on Philosophical Methodology*,
Philosophers in Depth,
https://doi.org/10.1007/978-3-030-02432-1_4

argument that metaphysics should be considered a historical science. Since this argument is not developed in *An Essay on Metaphysics*, the essay 'Function of Metaphysics in Civilization' is indispensable for a proper understanding of it. That, in Collingwood's view, metaphysics is a historical science has important implications. For this implies that it is purely descriptive by nature, without any assessment being involved. This position is explicitly endorsed by Collingwood, both in 'Function of Metaphysics in Civilization' and in *An Essay on Metaphysics*, though in the latter it is taken as starting point, while in the first it is substantiated.

It is understandable that Collingwood's position with respect to a major philosophical issue has met with little sympathy among philosophers. The background of this is that the nature of philosophical and historical discourses is quite different. For philosophers are used to speak about a subject in direct speech, while historians in speaking about the past usually speak in indirect speech (*oratio obliqua*). As explained by Frege, the difference between both forms of speech boils down to the circumstance that when in indirect speech a sentence incorporates a subordinate sentence referring to a thought, the latter does not have its truth-value as a reference, as is the case with direct speech, but the sense of the thought concerned. Since in Collingwood's view all history is the history of thought, it is obvious that indirect speech is for him just common practice, it accordingly being part of his re-enactment doctrine with the notion that past thought is incapsulated in present thought. To put it simply, one could write, for instance, a sensible history of flat-earth believers—that is, using indirect speech—while a discussion between scientists about the issue is another matter.

All the confusion that *An Essay on Metaphysics* has aroused has its origin in the difference between a purely philosophical and historical approach. Collingwood was probably less aware of this, he not only being both a philosopher and a historian, but also a philosopher of history. What for him was just common knowledge and practice has less been the case with philosophers, however. So the latter misinterpreted Collingwood's historical view on metaphysics in terms of direct speech, accordingly interpreting it as a form of 'extreme' historicism or scepticism. But Collingwood strongly advocated the historical approach, not only with respect to metaphysics and philosophy in general (see his chapter on 'The History of Philosophy' in his *An Autobiography*) but also in anthropology and human studies in general. He was right in this, since it is only through this approach that human conditions and experiences may be understood properly. In this connection it is noticeable that the historicism Collingwood has been blamed for by philosophers is generally considered a maxim by historians since its promulgation by the German historian Ranke, the 'father' of modern history, 200 years ago.

It is unfortunate, however, that Collingwood started his *Essay on Metaphysics* by expounding (in direct speech) his doctrine of absolute presuppositions based on the logic of question and answer. For this was not necessary, especially since in 'Function of Metaphysics in Civilization' he had explained his main argument that metaphysics is a historical science, speaking about metaphysical propositions instead of absolute presuppositions. In conclusion one could say, therefore, that the main argument of *An Essay on Metaphysics* has been misunderstood by its critics, while the debates about the doctrine of absolute presuppositions have only distracted attention from the main argument, accordingly arousing confusion about the nature of Collingwood's book about metaphysics.

This is especially unfortunate since Collingwood's main argument that metaphysics is a historical science is not only original but also thought-provoking. For history shows that humankind has indeed adhered to different metaphysical positions, which actually functioned as guiding principles for all aspects of life, both in thought and in action. These positions have the character of being self-evident and accordingly taken for granted. They are not substantiated, but simply presupposed. These metaphysical positions vary not only in history but also with respect to civilizations. Though certain metaphysical positions may be questioned, Collingwood argues that this is not the task of metaphysicians but of scientists in science, and activists in various realms of social life.

With his theory about the nature of metaphysics Collingwood has not only developed a cogent argument but has also clarified the importance of metaphysics as the unquestioned guiding principles of a particular time in a specific society or civilization. It is by becoming aware of these principles that the sciences have developed by questioning them, as slavery has been abolished and the position of women has changed through social pressure. According to Collingwood it is the task of metaphysicians to uncover the metaphysical positions involved, but it is the task of scientists and activists to change them. In the present world this issue is of more topical interest than ever.

2 Metaphysics in History: 'Function of Metaphysics in Civilization'

Collingwood wrote *An Essay on Metaphysics* in quite unusual circumstances: on board a ship to and from the Dutch East Indies on a voyage to this country from October 1938 to April 1939. In Hilary Term 1938 he had lectured on Central Problems of Metaphysics, and part of the manuscript concerned (pp. 29–52), with the title 'Function of Metaphysics in Civilization', has survived and has been added to the revised edition of *An Essay on Metaphysics* (EM: 379–421). The manuscript may therefore be considered a preparation to the book he started to write later that year.

The manuscript begins by referring to the role and importance of presuppositions, which he equates a little later with metaphysics, which he describes as "a presentation of the principles it discovers, not merely in the form of a catalogue, but in the form of a system" (EM: 383). "The essence of a metaphysical system is", he asserts, "that each item in it forms one among a number of presuppositions, all of which are actually presupposed by the thinking which the metaphysician is analysing. The presence of a given item in a metaphysical system is a question of fact, not a question of logic" (EM: 384).

It is of some importance to take notice of the position taken here by Collingwood. For in the discussions about the nature of Collingwood's view on metaphysics it has been a subject for debate if or in what sense various metaphysical positions might be assessed. In 'Function of Metaphysics in Civilization', however, Collingwood is quite clear about this issue when he maintains: "The question whether a pluralistic or a monistic metaphysics is *right* is a question which has meaning only when 'right' is taken to mean 'correct as a description of a certain subject matter'" (EM: 385–86). Comparing the metaphysics of Aristotle and Kant, Collingwood observes that "Aristotle was discovering by analysis the presuppositions of Greek science: Kant, those of modern and in particular Newtonian science" (EM: 387), which simply amounts to both philosophers having registered the scientific principles of their own times.

In an interesting passage, Collingwood raises the issue whether one kind of metaphysical system may be considered better than another, for instance a monistic one being preferable to a pluralistic one or vice versa. One could argue, says Collingwood, that one kind of metaphysical system is preferable because the science that is based on its presuppositions "may be a more adequate or powerful science than one constructed on presuppositions of another kind. For example: the substitution of monistic for pluralistic presuppositions in the science of nature of the Renaissance was surely a very great step forward" (EM: 390). The issue raised here by Collingwood is as important as it is challenging. It is also an issue under dispute among various interpreters of Collingwood's metaphysical views, for example, whether he was a relativist, or became a relativist and, if so, when. His views on the idea of progress, and those on the historical process in general are also involved in these debates. Thus it is important to quote the relevant passage in full. Collingwood starts his answer to the question that he himself has raised:

> Without challenging the statement that modern natural science is vastly better than the science of the Greeks and the Middle Ages, I will first remind you that in spite of its name metaphysics is not related only to natural science. It is related to every kind of systematic or orderly human thinking. If modern metaphysics is so much better than ancient, the improvement in natural science which is cited in evidence of this ought to be reinforced by a parallel improvement in all other branches of theoretical and practical thinking: e.g. in morals, in law, in politics, in economics, in religion, in art, etc. (EM: 391)

But Collingwood also puts forward an argument from another perspective. For he maintains that "deciding which of two things is the better is the same as choosing between them. In that case goodness is always relative to an act of choice, and where there is no act of choice between two things, to speak of one as better than the other is meaningless" (EM: 392). For this reason he declares that "[i]f anybody ... asks whether ancient or modern science is the better, I propose on this hypothesis to reply, not, 'I don't know,' but, 'the question is meaningless.' ... The question which kind of science is the better, therefore, is a nonsense question because to ask it presupposes the existence of a situation which does not exist" (EM: 393). Collingwood accordingly claims that it is meaningless to say of Aristotelian pluralistic metaphysics and the Kantian monistic one that "one is intrinsically better than the other, whether this means that, of Aristotle and Kant, one was a better metaphysician than the other, or that, of the systems of science resulting from the respective acceptance of these metaphysics, one system of science is better than the other" (EM: 393–94). He adds to this that "[t]he questions which these statements presuppose are questions which 'do not arise'", this being a phrase that is used in (A: 37–8) and (EM: 25–6) as well. "The question whether Greek science or modern science is the better can certainly be both asked and answered", Collingwood clarifies, "but only by way of a game, like the question 'which would you rather be, a cat or a dog,' or 'what would you wish for if a fairy gave you three wishes'" (EM: 394).

The next question he raises is "whether it is in *no* sense the business of the metaphysician to criticize or suggest improvements in the metaphysical presuppositions which he discovers by analysis to be implied in the thought of his community" (EM: 394). In this case there is no question of comparing the metaphysics of past science with modern science but of assessing the metaphysics of present science. But this suggestion is also rejected by Collingwood, after discussing various possible criteria to assess metaphysical presuppositions, viz. that they are self-contradictory, inherently improbable, or untrue (EM: 395–401).

Of particular importance is the argument used by Collingwood as regards his claim that with regard to metaphysical presuppositions their truth is not involved. He refers in this connection to the general principle that "[I]f *p* is an 'ordinary' (non-metaphysical) proposition, somebody's asserting *p* is one thing, and *p*'s being true is another thing, and there is no connection between these two things" (EM: 402). He refers here to *oratio obliqua* or indirect speech, as expounded by Frege in his famous article 'Über Sinn und Bedeutung' ('On Sense and Reference') of 1892. He there explains that when a sentence incorporates a subordinate sentence that refers to a thought, the latter does not have its truth-value as a reference, as is the case with normal sentences, but the sense of the thought concerned. He gives as an example the sentence 'Copernicus believed that the planetary orbits are circles'. In a case like this, Frege says, the truth of the main sentence is independent of the truth or falsehood of the subordinate sentence. In this case the main sentence is true, but the subordinate one is not (Frege 1997: 160).

Collingwood hints quite confusingly at the nature of indirect speech by stating that "*the propositions asserted by a metaphysician as such are not metaphysical propositions*", especially because he refers in both cases to propositions (EM: 404). On further consideration he seems to realize the confusion involved in speaking of both metaphysical propositions and the ones asserted by metaphysicians. Discussing the example that in our ordinary thinking we presuppose that all events have causes, he maintains that in this case a metaphysical error is only involved when people in their ordinary thinking in fact don't presuppose this, and not that in fact all events don't have causes (EM: 405). Collingwood then observes "that asserting a proposition is one thing and presupposing it quite another". In order to clarify this he maintains that "if people presuppose in their ordinary thinking that all events have causes, and arrange all their ordinary thinking in terms of that presupposition, it is difficult to see what could be meant by saying they have no right to make it", adding that "[s]uch a statement would be rather like saying that people have no right to measure lengths in feet and inches" (EM: 406).

What Collingwood's position amounts to is that on the one hand he implicitly refers to the nature of indirect speech (for with metaphysical presuppositions a thought is involved), where the truth-value of the main sentence is independent of the truth or falsehood of the subordinate sentence; but on the other hand denies that with metaphysical presuppositions the truth-value of the subordinate sentence is at issue, as in Frege's example of 'Copernicus believed that the planetary orbits are circles'.

So in Collingwood's view metaphysical presuppositions have a specific status in indirect speech in that they cannot be assessed. That is to say, in a metaphysical proposition the truth-value is confined to the main sentence, while the truth-value of the subordinate sentence is not involved. Accordingly Collingwood maintains that "the propositions asserted by metaphysicians ... are really statements of fact verifiable in the ordinary way in which all statements of fact are verifiable; i.e. according to historical principles" (EM: 407–8). He adds to this that "[s]ince ... the situation presupposed by the question 'is this proposition true or untrue?' does not exist in the case of a metaphysical proposition, that question as applied to a metaphysical proposition is a nonsense question". What we mean by saying that a metaphysical proposition is true, Collingwood argues, is "that we do actually presuppose it in our thinking", for example that every event must have a cause (EM: 409). When one discovers that other people use or have used a contradictory presupposition, this situation is described "by saying that their own presupposition is true and the contradictory one false". "Thus in the case of a metaphysical proposition", Collingwood concludes, "the statement 'this proposition is true' means 'I presuppose it', the statement 'this proposition is false' means 'I presuppose something incompatible with it'" (EM: 411). As regards the position of metaphysicians in this matter as compared to ordinary or scientific thinking, he maintains that with the latter metaphysical propositions are "neither asserted nor denied but merely presupposed", while "metaphysicians as such assert their truth or untruth. These assertions are purely historical assertions" (EM: 411). When he adds that "[t]o say that a given metaphysical proposition is true or untrue is thus the metaphysician's business and no one else's", this implies that metaphysicians actually are historians, or at least should behave as such (EM: 412).

This is, of course, a rather unorthodox position, to put it mildly, and many a philosopher will undoubtedly frown at encountering a viewpoint that at first glance sharply deviates from what philosophers usually consider metaphysics to be and the task of metaphysicians to be about. In order to assess Collingwood's position it is necessary to have a clear idea of its main elements. Since Collingwood's conclusion implies various interrelated aspects, it is necessary to quote it in full:

> The question whether it is a metaphysician's business merely to ascertain and describe historical facts (as I have said) or also to criticize the thought which he describes, is now answered. In his relation to the metaphysical propositions presupposed by the ordinary thinking of the community to which he belongs, the function of the metaphysician is purely historical or descriptive and in no sense critical. In his relation to the presuppositions of ordinary thinking in communities other than his own his function is again purely historical and in no way critical. In comparing these different systems of presuppositions, once more his function is merely historical. But in his relation to other metaphysicians his function is critical. If for example … metaphysicians have got into the habit of thinking that … [the science of his own day] presupposes the law of universal causation whereas really it does not, then his business is twofold: (*a*) to state the facts, (*b*) to criticize the metaphysicians who have misinterpreted them. (EM: 412–3)

Collingwood also emphasizes that metaphysical presuppositions are not constants but variables, both as regards civilizations and in history. "This is the lesson about metaphysical propositions which anybody *must* learn as soon as he begins seriously to study the history of thought", he says. In this respect he warns us to be careful in dealing with eighteenth-century metaphysics because in that time "metaphysicians of whatever school took it for granted that the metaphysical presuppositions they detected in European science of their own day were in fact common to all nations and to all periods of history" (EM: 419). Collingwood ends the manuscript with the plea that "[w]e must start again at the beginning and construct a new metaphysical theory which shall face the facts revealed by history. This has never yet been done" (EM: 420).

The reason to discuss the manuscript of 'Function of Metaphysics in Civilization' is that it throws much light on *An Essay on Metaphysics*. Collingwood wrote it in January 1938, and about ten months later he started to write *An Essay on Metaphysics*. There is therefore every reason to consider that 'Function of Metaphysics in Civilization' is related to *An Essay on Metaphysics*. The more so, since he ends the manuscript by expressing the need to construct a new metaphysical theory. Bearing in mind that the year in which he wrote his book on metaphysics began with a forerunner in the form of his *Autobiography*, it is clear that in *An Essay on Metaphysics* Collingwood aims at constructing himself the new metaphysical theory he envisaged. 'Function of Metaphysics in Civilization' is especially of interest for a better understanding of *An Essay on Metaphysics*, because it develops an argument that is not to be found in the book, and is there actually taken for granted.

Collingwood's claims that metaphysical presuppositions as such have no truth-value and accordingly cannot be judged or assessed. The only ones who are in the position and have the capability to assess metaphysical presuppositions are metaphysicians. But, and this is perhaps Collingwood's most startling position, they actually do not do this in their capacity as metaphysicians but as *historians*. That is, they must determine what are the metaphysical presuppositions in the present or have been in the past, without any value judgement playing a part. This is what Collingwood means by the 'historical method' of metaphysics, which is explained at length in 'Function of Metaphysics in Civilization'. In *An Essay on Metaphysics* it is frequently referred to as well, but not explained, with the result that in the debates on the book this central thesis about metaphysics has almost passed unnoticed, attention being focused instead on the notion of absolute presuppositions. I think Collingwood is to blame for this, because it was not necessary to develop this notion, which has only distracted attention from his main thesis that metaphysics is a historical science.

The manuscript of 'Function of Metaphysics in Civilization' is of interest for another reason as well. It was not meant to be published and was written in preparation for his lectures on metaphysics in Hilary Term 1938. The way it is written is of interest, since it shows how Collingwood grapples as it were with a problem he did not yet fully grasp. In the first part of the manuscript he rambles, for instance, from one subject to another, from the First World War to declaring that metaphysics is not only related to natural science but to all branches of theoretical and practical thinking, and that a metaphysical system is not an act of choice.

The manuscript could therefore be considered an attempt to 'think on paper', as he qualifies his 'Notes towards a Metaphysic', written in 1933–4 (PH: 119), though this manuscript is less systematic than 'Function of Metaphysics in Civilization'. For the latter develops an argument, ending with a clear conclusion of what in Collingwood's opinion metaphysics is about.

3 Metaphysics in History: *An Essay on Metaphysics*

In the Preface to *An Essay on Metaphysics* Collingwood gives a short explanation of its nature, saying that "[t]his is not so much a book of metaphysics as a book about metaphysics". In the opening chapters its characteristics are explained, developing his noted theory of absolute prepositions. His argument begins by stating that every statement is based on a question asked, and that every question in its turn involves a presupposition that is relative to it. Since in both cases a relationship is involved, he defines a relative presupposition as "one which stands relatively to one question as its presupposition and relatively to another question as its answer" (EM: 29). After this Collingwood defines an absolute presupposition as "one which stands, relatively to all questions to which it is related, as a presupposition, never as an answer" (EM: 31). He adds to this that not being propositions they are simply presupposed and not propounded (EM: 32–3). In his *Autobiography* the nature of absolute presuppositions is also described briefly, in reference to "my 'logic of question and answer'" (A: 66–7).

Though 'Function of Metaphysics in Civilization' and *An Essay on Metaphysics* deal with the same subject, there are two striking dissimilarities. For what in 'Function of Metaphysics in Civilization' are called metaphysical presuppositions are referred to in *Essay on Metaphysics* with the notion of absolute presuppositions. Besides this, where Collingwood *explains* the character of metaphysical presuppositions in 'Function of Metaphysics in Civilization' by *arguing* that they are historical statements, in *An Essay on Metaphysics* the character of absolute presuppositions is *demonstrated* within the context of the logic of question and answer. The strikingly different approaches to the same subject in a short period of time is easily explicable: 'Function of Metaphysics in Civilization' was written in preparation for lectures, while *An Essay on Metaphysics* was aimed at a general public and, not omitting, the positivistic opponents of metaphysics, who were at that time especially prominent and influential at

Oxford. But having written the manuscript Collingwood wrote *An Autobiography*, which was also written for a general public, its purpose being "to put on record some brief account of the work I have not yet been able to publish, in case I am not able to publish it in full" (A: 118). He made it clear in his *Autobiography* that the logic of question and answer played a prominent part in his intellectual development, it being based on his 'laboratory of knowledge' of archaeological practice and historical research (A: 24–5), transferring it therein to a philosophical notion as well. From archaeology to metaphysics is a rather uncommon trajectory, but, anyhow, that was Collingwood's. In this connection it is understandable that he cherished the logic of question and answer as his pet notion being suitable for archaeology, history, and philosophy, the three fields he worked on all his life.

It is infelicitous, however, that Collingwood developed his theory of absolute presuppositions within the context of his theory of the logic of question and answer. For this theory is no logic in the proper sense of the word and should in this case rather be taken figuratively. If the logic involved were to be taken seriously it is easily susceptible to criticism, as Michael Krausz has shown in his penetrating article 'The Logic of Absolute Presuppositions'. He argues that Collingwood's viewpoint that absolute presuppositions cannot give answers to questions leaves room for two possible positions: this is either the case in the *given* systematic inquiry or in *any* systematic inquiry (Krausz 1972: 223). Though one could of course object that, by dealing with history, by definition only given systematic inquiries are involved, this restriction is not made by Collingwood in his exposition of the nature of absolute presuppositions. Anyhow, Krausz argues that when Kant absolutely presupposed that a cause and its effect are related by a necessary connection and by way of a temporal sequence "neither part of the above conjunctive statement is presuppositionless; both presuppose that there is such a thing as a causal nexus with distinctly characterizable relations". For this reason Krausz concludes that "Collingwood adduces no presuppositionless example of an absolute presupposition" (ibid.: 238–9).

The notion of absolute presuppositions is discussed in an extensive and detailed article by Michael Beaney (2005). He argues that this notion could have been put forward by Collingwood independently of the logic of question and answer (Beaney 2005: 46–7, 70). I think he is right in this, for we have seen that its essence is already explained in 'Function of Metaphysics in Civilization' without using this notion. That Collingwood has used his logic of question and answer to substantiate the notion of absolute presuppositions is particularly unfortunate because it has greatly contributed to an improper understanding of *An Essay on Metaphysics*. Beaney exemplifies this when he maintains that "Collingwood's doctrine of absolute presuppositions is one of the most well-known features of his philosophy", and that "this doctrine was embedded in a conception of the nature and role of metaphysics" (ibid.: 41). And when he asserts in the same vein that "Collingwood's conception of presuppositions, as articulated in chapter 4 of *An Essay on Metaphysics*, 'On Presupposing' … is the key chapter of the book", while he speaks elsewhere of "the central claim of *EM*—that absolute presuppositions are neither true nor false",(ibid.: 43, 68) this position displays a misconception of the nature of *An Essay on Metaphysics*, confusing its goals and means. For it is hard to believe that when Collingwood ends 'Function of Metaphysics in Civilization' with saying that we must start with constructing a new metaphysical theory he had the doctrine of absolute presuppositions in mind.

Beaney may be right in claiming that the conception of absolute presuppositions is the most widely discussed aspect of *An Essay on Metaphysics*; indeed I suppose he is, but that does not imply that it is also the main argument of this book. In fact this is obviously not the case, since the thesis that metaphysics is a historical science is actually its central argument. Accordingly chapters VI ('Metaphysics an Historical Science') and VII ('The Reform of Metaphysics') are the key chapters, the latter title fitting in with the last sentence of 'Function of Metaphysics in Civilization', where Collingwood mentions constructing "a new metaphysical theory". The arguments developed in *An Essay on Metaphysics* are also the same as in the earlier manuscript, in particular as regards the thesis that metaphysics is a historical science. On the first page of chapter VI this thesis is again stated unambiguously: "All metaphysical questions are historical questions, and all metaphysical propositions are historical propositions", adding to this that "[t]his is the central point of the present essay" (EM: 49). So it is rather peculiar to suggest that Collingwood would be wrong about his book, and that it is actually about the theory of absolute presuppositions, the topics of metaphysics as a historical science and its role in history accordingly just being illustrations of this thesis.

Philosophers are of course not to blame in being primarily interested in philosophical questions and not in the trade of historians. But one cannot emphasize enough that Collingwood was as much a historian as a philosopher, as is evidenced by his *Autobiography*, in which both subjects are amply represented. The chapter on 'The History of Philosophy' is after 'Roman Britain', the largest chapter of the book, discussing some interesting topics, among other things that metaphysics can become a science "by becoming what it always has been; that is, frankly claiming its proper status as an historical inquiry" (A: 67), a claim which is phrased almost identically in (EM: 77). The point I want to make is that the only way to grasp the nature of the latter book it is to take seriously Collingwood's claim that metaphysics is a historical science.

This means that certain principles and characteristics of historical thinking are involved. In historical studies certain descriptions, explanations, and narratives are given of the past. In doing this they characteristically incorporate indirect speech, that is—as explained above—when a main sentence incorporates a subordinate one referring to a thought, only the main sentence has a truth-value while the subordinate sentence only has a sense. In indirect speech the notion of thought should be taken in a broad sense, implying words like 'say', 'being convinced', 'believe', 'conclude', and so on. Collingwood's well-known thesis that all history is the history of thought therefore implies that historians typically use indirect language. That for Collingwood, as regards this aspect, there is indeed a historical principle involved is aptly demonstrated in the following passage in *The Principles of History*:

> All history is the history of thought; and when an historian says that a man is in a situation this is the same as saying that he thinks he is in a situation. The hard facts of the situation, which it is so important for him to face, are the hard facts of the way in which he conceives the situation. If the reason why it is hard for a man to cross the mountains is because he is frightened of the devils in them, it is folly for the historian, preaching at him across a gulf of centuries, to say 'This is sheer superstition. There are no devils at all ...' ... Sheer superstition, no doubt: but this superstition is a fact, and the crucial fact in the situation we are considering. The man who suffers for it when he tries to cross the mountains is not suffering merely for the sins of his fathers who taught him to believe in devils, if that is a sin; he is suffering because he has accepted the belief, because he has shared the sin. If the modern historian believes that there are no devils in the mountains, that too is only a belief he has accepted in precisely the same way. (PH: 100–1; IH: 317–8)[1]

Collingwood wrote this passage in February 1939 in the Dutch East Indies, having stopped working on *An Essay on Metaphysics*. In his log of his trip to the Dutch East Indies, he writes on 31 October 1938: "Writing all day, finished ch. 6" (Collingwood 2013: 460). This chapter being on 'Metaphysics an Historical Science', it is reasonable to suppose that the passage quoted actually is meant to be in line with the argument developed there, though in the first instance it looks quite otherwise. For referring to the metaphysical propositions involved in the Newtonian, Kantian, and Einsteinian schools of thought in physics, they being respectively that some, all, and no events have causes, Collingwood maintains that "[e]ach [school of physical science] is important, and fundamentally important, to the science that makes it, because it determines the entire structure of that science by determining the questions that arise in it, and therefore determining the possible answers. Thus every detail in these respective sciences depends on what absolute presuppositions they respectively make" (EM: 52). He thereupon says one can do two things with absolute presuppositions: what the scientist does and what the metaphysician does. In this respect the business of the scientist is to presuppose the absolute presuppositions concerned (EM: 54), while "[i]t is the proper business of a metaphysician to answer the question what absolute presuppositions are or were made by Newtonians, Kantians, Einsteinians, and so forth. These are historical questions" (EM: 55).

There is of course an air of absurdity in a comparison of Collingwood's argument about the various absolute presuppositions involved in the history of science with the beliefs of devil-fearers he wrote about a few months later. It is nonetheless clear that in both accounts the same issue is involved, though the observation should be made that in the case of the devil-fearers Collingwood should have referred to anthropologists instead of historians. But that is not relevant for the argument involved, because in both descriptions reference is made to implicit thoughts, in the case of scientists called absolute presuppositions, and of the devil-fearers beliefs. Besides this, in both of them indirect speech is involved, implying that in the subordinated sentence referring to a thought or belief only its sense is at issue and not its truth-value.

In their *Research Companion to Collingwood* the editors make an interesting observation concerning Collingwood's development and interests in the late 1930s, saying of the year 1936 that "[a]t this time Collingwood was becoming increasingly interested in anthropology and folklore … In the light of this it can be seen that his work on the philosophy of history, Roman Britain, civilisation, folklore, the philosophy of art, mind and the emotions, and metaphysics as the search for absolute presuppositions forms an interconnected whole" (Connelly et al. 2015: 47). I think this assessment is correct and is evidenced by Collingwood's publications from *Roman Britain and the English Settlements* of 1936 to *The New Leviathan* of 1942. Though there is a great variety of subjects dealt with by Collingwood during this period, there are indeed unarguably interconnections between them.

This also holds true for the important and highly interesting manuscripts about folktales and anthropology, which were published in 2005. Especially interesting is Collingwood's extensive criticism of the way 'primitive' people or 'savages' have been dealt with in the past (PE: 132–77). He makes a chronological distinction between the three interpretative phases of the philologists (Grimm and Max Müller, 1810–70), the functional anthropologists (Tylor and James Fraser, 1870–1910), and the psychologists (Freud and Jung). Collingwood criticizes all three approaches for their naturalism, that is, "[e]ach of them treats its subject matter as something to be contemplated from without, something external to the thinker, something that is not himself but something else" (PE: 181). Historical naturalism is a recurrent theme in Collingwood's philosophy of history, and is dealt with, for instance, in 'Human Nature and Human History' (IH: 205–31), and extensively in *The Principles of History* in a chapter on 'Nature and Action' (PH: 78–110). But it is interesting to see how he opposes naturalism within the context of anthropological theories. For the approach he advocates is developed in a chapter under the intriguing but telling title 'The Historical Method' (PE: 178–94). Collingwood declares that this position is based on two maxims, one of Spinoza and one of Bishop Butler. That of the first is, Collingwood says, "neither to condemn nor to deride the feelings and actions of men, but to understand them" (PE: 184),[2] while Bishop Butler's maxim is "that every thing is what it is, and not another thing" (PE: 186),[3] this implying, Collingwood asserts, "as a rule of method … formally to recognize that our study is historical, not naturalistic, in character". This implies, he observes hereafter, "that man is man and not merely nature, thinker and not merely object thought about … The only way in which man's thought and actions can become the object of human knowledge is through the methods of history" (PE: 193).

Collingwood wrote these words two years before working on *An Essay on Metaphysics*. In the first instance it seems rather far-fetched to see a relationship between his manuscripts on anthropology and his theory of metaphysics as historical study. After all, anthropologists don't speak of their trade as being historical. The point Collingwood makes, however, is that he advocates the historical *method* as being the proper anthropological approach, which is not the same as the *study* of history as the study of the past. So, in Collingwood's argument in anthropology only the historical method is involved, while in metaphysics both the study of history and its methods are involved. Accordingly, there is indeed a *methodological* likeness between them, and this is evidenced by the way Collingwood illustrates his claim that anthropology should use the historical method:

> Anthropology … is a historical science, where by calling it historical as opposed to naturalistic I mean that its true method is thus to get inside its object or recreate its object inside itself. Anyone can see this if he considers, for example, the historical study of early mathematics. Greek geometry, and Egyptian geometry even more, is a crude and primitive thing compared with the geometry of our own day; but if we are to study it historically (and there is no other way of studying it) we must examine its documents until we find ourselves able to think again for ourselves the thoughts of those early geometers. There is no difference of principle between this and any other case in which an attempt is made to study the early history of human ideas. The universal rule in every such case is to reconstruct for oneself and in oneself the ideas whose history one is trying to study. If one cannot do this, one may speak with the tongues of men and of angels, but the result is not anthropology. (PE: 153–4)

So we see Collingwood here giving a historical example in an argument about the method to be used in anthropology, while we have seen that in *The Principles of History* he used the anthropological example of the devil-fearers in an argument about a principle of historical methodology. But it is clear that in either case the historical methodological issue involved is Collingwood's basic assumption.

4 *An Essay on Metaphysics* Under Discussion

It is an understatement, to put it mildly, that Collingwood's thesis of the nature of metaphysics as explained in *An Essay on Metaphysics* has not been generally endorsed. For it has aroused much debate and has been the target of various criticisms. Leaving the notion of absolute presuppositions aside, Collingwood has especially been blamed for the fact that his theory of metaphysics implies a too pronounced relativism with respect to this subject, or, as it is sometimes phrased, his theory represents an extreme form of historicism. Knox has set the tone in his preface to the first edition of *The Idea of History*, being outrightly negative about *An Essay on*

Metaphysics, saying it expresses a philosophical scepticism that is bound up with a new dogmatism (IH: 1st ed., xvi).[4] Later comments on the book have been more balanced, but the subject of relativism frequently turns up, for instance in the article with the telling title 'Conceptual Change and the Problem of Relativity' by Stephen Toulmin (1972a). He raises the question "how can we retain what is sound in Collingwood's account, without being driven all the way down the road to historical relativism?" (Toulmin 1972a: 214). His suggestion is that "historical relativity might be tolerated on the theoretical level, provided only that the relevant disciplinary considerations were historically invariant" (ibid.: 217). In the same vein, in *Human Understanding* (1972b), Toulmin introduces Collingwood as a relativist, who yields entirely to the diversity of conceptual systems, and "abandons the attempt to judge impartially between different cultures or epochs, and treats the notion of 'rationality' as having no more than a local, temporary application" (Toulmin 1972b: 53). Michael Krausz for his part qualifies Collingwood's theory of absolute presuppositions as implying "an extreme relativism which rules out the possibility that an inquirer might offer objective reasons for abandoning one inquiry in favour of another. Consequently, the historian of ideas could account for such change only in non-rational terms" (Krausz 1972: 222).

Rex Martin has discussed on various occasions Collingwood's view on metaphysics as expounded in *An Essay on Metaphysics*, in which the issue of his supposed relativism plays a prominent part (Martin 1981, 1995, 1998, 2013).[5] In his editor's introduction to the revised edition of *An Essay on Metaphysics* he refers to Knox's and Donagan's view that between *An Essay on Philosophical Method* of 1933 and *An Essay on Metaphysics* of 1940 there is an obvious discontinuity as regards the nature of philosophy in that the differentiation between history and philosophy is abolished, "with the result that philosophy now became prey to the demons of scepticism and relativism". But Martin declares that his account "is strikingly different. Rather than a doctrinal break between Collingwood's two treatises on metaphysics, I see a continuity", he avers (Martin 1998: liv). Elsewhere he asserts that "Collingwood's strong endorsement of progress in science ... should indicate that he is not committed to either relativism or historicism in any negative or deleterious way" (Martin 2013: 362–3).

I would argue, however, that the way Martin develops his argument is open to serious criticism. An obvious example, and also the most crucial one, is when he declares that "Collingwood did not regard the outcome of this metaphysical project [of EM] to be merely reportorial. Rather,

Collingwood envisioned an element of critical assessment as part of the proper task of metaphysics" (ibid.: 354). But we have seen that Collingwood declares in 'Function of Metaphysics in Civilization' that "the function of the metaphysician is purely historical or descriptive and in no sense critical" (EM: 412), which is also explicitly stated in *An Essay on Metaphysics*.[6] Nevertheless Martin maintains that "I would be reluctant to regard Collingwoodian metaphysical statements … as essentially *descriptive* in nature" (Martin 2013: 354). But that is anyhow the way it is, viz. the way Collingwood conceived it and explained in *An Essay on Metaphysics*; that is, what he calls absolute presuppositions should be considered historical phenomena, which accordingly should be studied as historians are used to study the past, implying certain rules and principles. This is the 'historical method' that Collingwood time and again advocates in *An Essay on Metaphysics*, and is in accordance with Bishop Butler's maxim "that everything is what it is, and not another thing", as put forward by Collingwood in his manuscripts on anthropology.[7]

Martin, however, develops an argument that is at variance with this explicitly expressed view by Collingwood. It is based on two claims: that metaphysics is 'criteriological' and that absolute presuppositions are propositions (Martin 2013: 354, 362). The first claim is quoted out of context, however, since the notion of thought being criteriological is *only* developed by Collingwood with reference to the claim of psychology to be the science of thought and mind in general. He does it at various places, not only in (EM: 106–42) but also in (PA: 171n.) and in (PH: 84–8, 108). The passage in *An Essay on Metaphysics* Martin refers to in order to support his claim that metaphysics is 'criteriological' reads: "My suspicions … are about the status of psychology as the pseudo-science of thought which claims to usurp the field of logic and ethics in all their various branches, including political science, aesthetics, economics, and whatever other criteriological sciences there may be, and finally of metaphysics" (EM: 142). Since Collingwood considers metaphysics to be a historical science it is obvious that this implies that it is criteriological, all the more so because in (PA: 171n) Collingwood explicitly mentions history as being criteriological.[8] So when Collingwood refers in (EM: 142) to metaphysics being criteriological, it is abundantly clear that this is the case because it is a historical science, which is put forward *ad nauseam* in *An Essay on Metaphysics*.

Martin's curious claim that absolute presuppositions as conceived by Collingwood are in fact propositions is based on the fact that he referred to them in *An Autobiography* as beliefs (A: 66–7).[9] But beliefs are propositional attitudes, Martin argues, and therefore not meaningful. Collingwood, however, speaks of the sameness of suppositions and propositions, he says, referring to a manuscript entitled 'Notes for an Essay on Logic' (EM: 422–7) that he wrote in 1939 (Martin 2013: 361). But what Collingwood writes in these Notes exactly indicates that he is correct with respect to the logical status of absolute presuppositions and 'beliefs'. For he speaks of *indirect* speech, saying "Phrases beginning 'that' (the *oratio obliqua* [or indirect speech] formula) may just as easily express suppositions as propositions" (EM: 423). So that is the passage Martin refers to when he speaks of the sameness of content as between suppositions and propositions, disregarding that Collingwood speaks of indirect speech and not direct speech. It is therefore not without reason that in the foregoing the importance has been emphasized of realizing that the use of indirect speech is crucial with respect to a proper understanding of Collingwood's theory of metaphysics being a historical science, the use of indirect speech being characteristic of the latter. Frege gives the following example of the nature of indirect speech:

> If, toward the end of the battle of Waterloo, Wellington was glad that the Prussians were coming, the basis for his joy was a conviction. Had he been deceived, he would have been no less pleased so long as his illusion lasted; and before he became so convinced he could not have been pleased that the Prussians were coming—even though in fact they might have been already approaching. (Frege 1997: 161)

It would in this connection be rather extravagant to say that these statements about Wellington have no meaning because convictions are propositional attitudes. But this is what Martin's argument boils down to.

Another controversial issue raised by Martin concerns the notion of progress. In this connection he refers to a passage in *The Idea of History* where Collingwood argues that "If thought in its first phase, after solving the initial problems of that phase, is then, through solving these, brought up against others which defeat it; and if the second solves these further problems without losing its hold on the solution of the first, so that there is gain without any corresponding loss, then there is progress" (IH: 329). In this connection Collingwood observes that "[t]he interest of science, in relation to the conception of progress, seems to be that this is the simplest and most obvious case in which progress exists and is verifiable", giving Newton's law of gravitation and that of Einstein as example. But, he adds, "science is and can be mistress only in her own house, and forms of activity which cannot progress ... cannot be made to do so by subjecting them, if that phrase meant anything, to the rule of science" (IH: 332).

It should be noted, in this connection, that Collingwood's view on the idea of progress is balanced and variegated, being dependent on a point of view, meaningless, meaningful, or necessary (Van der Dussen 2016: 124–31). He asserts, for instance, that no progress is involved in ways of life; in happiness, comfort, or satisfaction; in art; and in morality and economics only with respect to social institutions (IH: 326, 330–1). These are all areas in which absolute presuppositions are involved. But Martin focuses only on science, concluding from Collingwood's statement that with respect to the conception of progress science "seems to be ... the simplest and most obvious case in which progress exists and is verifiable", that he exhibited a "strong endorsement of progress in science" (Martin 2013: 362).[10] He not only disregards Collingwood's caveat that "science is and can be mistress only in her own house" but also that in *An Essay on Metaphysics* the notion of science is used in a broad sense, that is "not natural science alone but orderly and systematic thinking on every subject", and "not orderly and systematic 'theoretical' thinking alone but orderly and systematic 'practical' thinking as well" (EM: 85).

Speaking of scientific progress in absolute terms, Martin subsequently develops an argument in which Collingwood's theory of metaphysics is conceived as exhibiting a progressive development. This is done by relating the notions of a scale of forms and overlap of philosophical classes as developed by Collingwood in *An Essay on Philosophical Method* to the notion of absolute presuppositions in *An Essay on Metaphysics*. For this reason Martin suggests that absolute presuppositions could be interpreted as exhibiting a 'modified' scale of forms being appropriate to historical processes of cumulative and progressive development (Martin 2013: 366–9). When he says that Collingwood himself considered the sciences of history and archaeology not only to be progressive but also made a contribution to them, he is correct, of course (ibid.: 364). But he did this as a historian and archaeologist, as Einstein did in physics as a scientist, and not as a metaphysician.

But what about Collingwood's sarcastic comment on the First World War, saying that "[t]he War was an unprecedented triumph for natural science. Bacon had promised that knowledge would be power, and power it was: power to destroy the bodies and souls of men more rapidly than had ever been done by human agency before" (A: 90). This means that even if Collingwood would have considered scientific progress a model for the notion of cumulative and progressive development in general, as Martin argues, one should always keep in mind that with the development of absolute presuppositions much more is involved than the ones of science, and that, as noted above, in Collingwood's view also 'orderly and systematic thinking on every subject' and 'practical' thinking are involved.

Martin's argument that Collingwood's theory of metaphysics being a historical science should be interpreted in the sense of history as a cumulative and progressive development is based on the fact that in his opinion re-enactment cannot be involved in Collingwood's account of metaphysics. If this would be a correct assessment he would be right, and at least there would indeed be a problem. On this issue Martin declares that "Collingwood consistently regarded re-enactment as involved with actions. Specifically, the aim of re-enactment was to understand or explain an action by reference to the thoughts ... of an agent" (Martin 1998: xxxi). For this reason he argues that "unless absolute presuppositions are actions or can be the more or less direct effects of actions (and thereby the *intended* effect of action), there is little point in making re-enactment the focal point in any Collingwoodian assimilation of metaphysics to history" (ibid.: xxxii–iii).

Collingwood's theory of re-enactment of the past is the most widely discussed aspect of his philosophy of history and cannot be dealt here at length. In his *Autobiography* Collingwood explains the main characteristics of his "new conception of history" developed in 1928. It consists of three principles: (1) "all history is the history of thought"; (2) "historical knowledge is the re-enactment in the historian's mind of the thought whose history he is studying"; (3) "Historical knowledge is the re-enactment of a past thought incapsulated in a context of present thoughts which, by contradicting it, confine it to a plane different from theirs" (A: 110, 112, 114). In his manuscript 'Outlines of a Philosophy of History' (IH: 426–96), in which, among other things, the notion of re-enactment of the past is developed, Collingwood gives the following clarification of its nature:

> [T]o re-enact the past in the present is to re-enact it in a context which gives it a new quality. This context is the negation of the past itself. Thus, the historian of poetry, reading Dante, re-enacts the medieval experience which that poem expresses: but while doing this he remains himself: he remains a modern man, not a medieval: and this means that the medievalism of Dante, while genuinely revived and re-experienced within his mind, is accompanied

> by a whole world of fundamentally non-medieval habits and ideas, which balance it and hold it in check and prevent it from ever occupying the whole field of vision. For Dante, the *Commedia* was his whole world. For me, the *Commedia* is at most half my world, the other half being all those things in me which prevent me from literally becoming Dante. (IH: 447)

In this passage all three elements of the re-enactment doctrine are included: thoughts are implied in the medieval experience of Dante, these thoughts are re-enacted and as such incapsulated in present thought. It is in particular the feature of incapsulation that is crucial, though it is, strangely enough, mostly disregarded in the discussions on the re-enactment doctrine. The notion of incapsulation of past thoughts is, in particular, of importance since it epitomizes the notion of indirect speech, which is, as argued above, characteristic of historical studies. It is most obviously practised in literary history, intellectual history, religious history, the history of history (historiography), history of philosophy, and so on but also metaphysics as conceived by Collingwood, that is, as a historical science.

So Martin is off the mark when he claims that Collingwood consistently regarded re-enactment as being involved with actions. Though it is true, of course, that Collingwood is focused on the thought-side of actions, this is by no means the only occasion that re-enactment is involved. For this doctrine indeed covers a broad spectrum, Collingwood saying that "[i]f it is by historical thinking that we re-think and so rediscover the thought of Hammurabi or Solon, it is in the same way that we discover the thought of a friend who writes us a letter, or a stranger who crosses the street" (IH: 219). But it is telling that Collingwood speaks here of discovering the thoughts of a friend in a letter and someone crossing the street, but of rediscovering the thought of Hammurabi or Solon. For the difference is obvious, the thoughts of the first being directly re-thought, while those of the past have to be reconstructed. On repeated occasions Collingwood declares that historical knowledge is achieved through inferential reasoning based on evidence (IH: 133, 176, 234, 237, 251–2, 435; PH: 76).

Since Collingwood considers the study of metaphysics to be historical, this principle is involved as well. In this case the evidence is based on the writings of, say, Aristotle, Plato, Descartes, Locke, Kant, or Mill. Though this is just a commonplace, Collingwood once refers to it, saying "history … demands proof; and … the only way to prove that somebody has made or has not made a certain absolute presupposition is to analyse the records of his thought and find out" (EM: 60). That is, the thoughts incorporated in the records have to be re-enacted in being incapsulated in the present thoughts of the metaphysician/historian, as in Collingwood's example of Dante. Subsequently through inferential reasoning these incapsulated thoughts are used by the metaphysician/historian in his or her present thoughts as evidence for developing an argument about the nature of the metaphysics of the past concerned. As compared with the rethinking of the thoughts implied in actions, this procedure is indeed rather complicated, but it not only involves re-enactment but is also in line with Collingwood's philosophy of history in general. Martin is therefore right in saying that absolute presuppositions cannot be re-enacted in the way the thought-side of actions may be discerned. But they can be known through a procedure in which re-enactment is involved as regards the evidence (the records of the relevant thoughts) on the basis of which an inferential argument is made about particular absolute presuppositions.

Though Collingwood is known primarily for his theories in the field of the study of the past, he also developed specific views on the nature of historical processes and development. In (A: 97–8, 140–1), for instance, he briefly discusses the characteristics of historical processes, but his ideas on the subject are spread over the rest of his works as well.[11] A topic he discusses time and again is the difference between natural and historical processes, for instance in 'Human Nature and Human History' (IH: 205–31) and 'Reality as History' (PH: 170–208). With respect to historical processes it is of importance to realize that the notion of history is ambiguous in the sense that it may be used in two ways: as referring to the past itself (as with processes) or the study of it. The relationship between the two has traditionally been a major issue in the theory and philosophy of history. In the Preface of 'Outlines of a Philosophy of History', written in 1928, Collingwood declares that "this problem is discussed and solved by the doctrine of the ideality of history: for that doctrine lays it down that historical fact, as known to the historian, is essentially relative to the thought that knows it" (IH: 429). In this way the supposed gap between the historical 'facts' and the historian's thought about them is neutralized. This position is clarified by Collingwood as follows:

> [T]he necessary forms and conditions of historical thought are now seen to determine the necessary forms and conditions of its object. Everything that is said about history *a parte subjecti* can therefore be repeated, *mutatis mutandis*, about history *a parte objecti*. ... The gulf which, on an empiricist or positivistic philosophy, separates historical thought from historical fact, has disappeared. (IH: 429)

It should be noted that when Collingwood speaks of 'historical facts' he refers to 'historical reality' (history *a parte objecti*), and this includes historical processes. The latter play an important part in his philosophy of history, to the extent that "the essence of history lies not in its consisting of individual facts, however valuable these facts may be, but in the process or development leading from one to another" (IH: 169). Likewise, he criticizes Dilthey's philosophy of history, "because the individuals of which he was thinking were conceived as isolated past facts and were not integrated into a genuine process of historical development" (IH: 172). It is therefore of importance to realize that Collingwood's philosophy of history applies and is related to both individual historical 'facts' and historical processes that are seen as 'facts' as well, though of another nature. In *An Essay on Metaphysics* he does both with respect to absolute presuppositions, determining those of Newton or Kant, for instance. But in ch. VII ('The Reform of Metaphysics') he pays special attention to the process-view as regards the notion of a 'constellation' of absolute presuppositions. The relevance of the two approaches in the study of metaphysics is phrased by Collingwood as follows:

> To trace the process by which one historical phase turns into the next is the business of every historian who concerns himself with that phase. The metaphysician's business, therefore, when he has identified several different constellations of absolute presuppositions, is not only to study their likenesses and unlikenesses but also to find out on what occasions and by what processes one of them has turned into another. This is the only legitimate (that is, historical) way in which he, or anybody else, can answer the question 'Why did such and such people at such and such a time make such and such absolute presuppositions?' (EM: 73)

Martin is therefore on the wrong track when he says, referring to Collingwood's statement "Where there is no strain there is no history" (EM: 75), that "it amounts to a virtual equation of history with change" (Martin 1998: xxxviii). This is a rather curious observation, it being a truism that history is by its very nature a process of continuing changes. But Martin takes it as providing "support, from within the *Essay on Metaphysics* itself, for my argument … that a historical process of development is principally what Collingwood had in view when he said that metaphysics and its subject matter are historical" (ibid.). This implies that Martin is of the opinion that the study of historical processes and developments is quite another subject than the one of individual facts. But this is not only at variance, as indicated above, with Collingwood's explicitly stated views but also with historical practice, which on almost all occasions pays attention to both individual facts and developments or processes. As both apply to metaphysics in *An Essay on Metaphysics* as well, Collingwood accordingly is in line with customary historical practice.

5 Conclusion

Taking the foregoing comments on *An Essay on Metaphysics* into account, it is clear that Collingwood's theory of absolute presuppositions and his thesis of metaphysics being a historical science have not been generally approved of, to put it mildly. For even when one endorses Collingwood's claim that metaphysics is a historical science, which I am inclined to do, it actually cannot be done in all conscience without being more or less familiar with the multifaceted aspects of Collingwood's philosophy of history, which have to be reconstructed from bits and pieces.

The Principles of History, Collingwood's last work on the subject and written during his work on *An Essay on Metaphysics*, was described in a letter to his friend Gerald Simpson as "the book which my whole life has been spent in preparing to write" (Van der Dussen 2012: 363, 1981: 443–4). But Collingwood never finished his final book on philosophy of history, with the result that his central thesis of metaphysics being a historical science lacks a solid basis, since no explanation is given of what this science and in particular its principles are like. Phrases like "any one who is reasonably well acquainted with historical work knows that …" (EM: 66), which abound in *An Essay on Metaphysics*, are, of course, not very convincing, especially since there is no reason to expect that philosophers, for whom the book is primarily written, would all be acquainted with historical work, in particular as conceived by Collingwood. Apart from this, it is not surprising that philosophers will only be convinced by philosophical arguments and not historical ones, as is usually the same, one should add, with historians with respect to philosophers. But Collingwood was in the rather unique circumstance that he was not only a philosopher but also a prominent archaeologist and historian of Roman Britain. Because of this he considered the principles involved in the latter fields to be obvious, without realizing that they would be less obvious to philosophers. He could probably have prevented this by finishing and publishing *The Principles of History* before *An Essay on Metaphysics*.

But since this has not been the case, it is not surprising that *An Essay on Metaphysics* has been widely criticized. For the book has indeed some flaws. The theory of absolute presuppositions, for instance, was not necessary for its main thesis that metaphysics is a historical science, and we have seen that Collingwood had already developed it in 'Function of Metaphysics in Civilization' without using this notion. But it has in particular been this theory that has been criticized, with the result that Collingwood's main thesis sometimes disappeared from sight. Collingwood's thesis that metaphysics is a historical science has been hardly taken seriously, seen as a sign of a radical conversion of his previous views implying an extreme form of relativism, while Martin has developed an alternative view that is not in line with Collingwood's. This tendency has developed because Collingwood in fact did not so much argue that metaphysics is a historical science but rather took it for granted. Accordingly, in *An Essay on Metaphysics* it is a starting point instead of being presented as the conclusion of an argument. In his manuscript 'Function of Metaphysics in Civilization', however, he developed a serious argument with respect to his main thesis, at least in the last part of it, and he should accordingly have used it for *An Essay on Metaphysics.* Collingwood moreover too often digressed from his main argument by challenging adversaries of his views like the positivists of his time and psychology.

Insofar as Collingwood's thesis of metaphysics being a historical science has been taken seriously, most comments have been critical as well, the main argument being that it implies an extreme form of relativism or historicism. In conclusion I will argue that this position shows a defective notion of the study of history, and that accordingly the criticism levelled against Collingwood with respect to this issue is misplaced.

The position of historicism or historism (translation of the German *Historismus*)[12] is a viewpoint on the nature of history that was particularly prominent in Germany in the nineteenth century in reaction to the ideas of the Enlightenment. As regards history these ideas are epitomized in Hume's saying that "Mankind are so much the same, in all times and places, that history informs us of nothing new or strange in this particular" (Hume 1963: 65). It was especially the renowned historian Leopold von Ranke (1795–1886) who developed views that are at odds with Hume's position. To date Ranke evokes respect, especially as regards the new methods he introduced in the study of history, being the first to practise and advocate the systematic study of historical documents. John Barker maintains that "[t]he most fleeting exposure to historical studies today will show that Ranke's aims and standards

... have ruled nearly all professional historians ever since ... his principles of research have given the modern West its most characteristic notion of what history is" (Barker 1982: 159). Likewise, Ernst Breisach speaks of Ranke as being "widely celebrated as the 'father of historical science'", and that "German historiography had rushed ahead of the rest of European historiography in methodological sophistication" (Breisach 1983: 232, 263). For his part, the Dutch historian Pieter Geyl—otherwise being quite critical, for example, of Arnold Toynbee—says of Ranke that "[h]is achievement commanded universal respect. There is among writers of history hardly a second figure who is equally awe-inspiring" (Geyl 1974: 10).

Ranke has also set the tone for two theoretical historical principles. The first opposes the generally accepted view of the past that had been operative from antiquity to well into the eighteenth century, namely that history should have a didactic function, phrased in Cicero's well-known dictum that history is *magistra vitae* ('life's teacher'). Bolingbroke was in 1752 the last to express this position explicitly, saying "We ought always to keep in mind, that history is philosophy teaching by examples how to conduct ourselves in all the situations of private and public life" (Bolingbroke 1752: 18). Ranke, however, advocating a neutral and value-free approach to the study of the past, opposed the view of the past that had been current for almost two millennia, declaring in his first book: "To history has been assigned the office of judging the past, of instructing the present for the benefit of future ages. To such high offices this work does not aspire: It wants only to show what actually happened" (in Stern 1956: 57).[13] This phrase is undoubtedly the most quoted passage in the theory and philosophy of history, in particular the last words '*wie es eigentlich gewesen*', and Collingwood also referred to it (IH: 130, 393–4). But in an equally important passage which is less known, another important position is expressed by Ranke, for him a historical one, but in fact philosophical. It concerns the idea of progress, it being prevalent in the nineteenth century. Ranke strongly opposed this idea, since in his view it would be 'a divine injustice' ('*eine Ungerechtigkeit der Gottheit*') if a particular age were to have no value in itself, but only serve as an entry to a next and superior stage. He explains his position as follows:

> Such a quasi-mediatized generation would, considered by itself, not have a meaning; it would only mean something insofar as being the stage for the next generation, and would not be related directly to the divine. I maintain, however: each epoch is immediate to God, and its value in no way depends on what results from it, but in its existence itself, in its own self (translation mine).[14]

Both positions of Ranke have been accepted as a matter of course by all historians, and they could accordingly be considered absolute presuppositions of their field. In this connection Stern speaks of "Ranke's conception of history, which a later generation called historicism, and which has remained the dominant force in historiography" (Stern 1956: 54). So it is no coincidence that when Collingwood in his anthropological manuscripts explains the principles involved in the 'historical method', they correspond to the ones put forward by Ranke: Bishop Butler's maxim "that everything is what it is, and not another thing" equates with Ranke's '*wie es eigentlich gewesen*', while Spinoza's maxim "neither to condemn nor to deride the feelings and actions of men, but to understand them" equates with Ranke's dictum '*jede Epoche ist unmittelbar zu Gott*'. As regards the latter, historians after all no longer refer to the Middle Ages as the Dark Ages, nor do anthropologists refer to other cultures or people as 'primitive'. Taking this into consideration, one can only conclude that the critics, who blame Collingwood for the 'danger' of historicism and relativism implied by his theory of metaphysics as a historical science, are simply ignorant of some of the most fundamental principles of the study of history. But Collingwood was not, and it is therefore no coincidence that the two principles he put forward in his manuscripts on anthropology as maxims of the 'historical method' equate to those of Ranke.

The issue of Collingwood's supposedly radical conversion to a historicist and relativistic position, that started with Knox's judgement, may also be seen from another perspective. In this connection Connelly has in a balanced and convincing article argued that the 'radical conversion' argument as regards Collingwood's development is defective. His observation that Collingwood's statement that "philosophy as a separate discipline is liquidated by being converted into history" (IH: 1st. ed., x) should be seen as an overlap between philosophy and history as explained in *An Essay on Philosophical Method* is of especial interest (Connelly 2013: 408).

But I would put this issue differently and more succinctly by maintaining that *An Essay on Philosophical Method* is a philosophical study, while *An Essay on Metaphysics* in its argument is primarily a historical study. The first four chapters of it, however, dealing with the notion of absolute presuppositions, are of a philosophical nature. This implies an ambiguity in the framework of the book, inciting controversies and misunderstandings. To put it otherwise, in the first four chapters of *An Essay on Metaphysics* Collingwood develops in direct speech his theory of absolute presuppositions as a philosopher, while he deals in the rest of the book as a historian with the same subject in the mode of indirect speech about historical practice that in its turn also uses indirect speech. Being both a philosopher and historian, it is understandable that he did not consider both endeavours to be incompatible, but it is also understandable that most of his readers thought otherwise. *An Essay on Metaphysics* may be considered exemplary of the *rapprochement* between philosophy and history, as advocated in (A: 77). This has not been considered generally as such by its readers, however, who tend to be philosophers, failing to realize the importance and impact of the historical line of approach as advocated by Collingwood in *An Essay on Metaphysics.*

On another occasion I have observed that "it is noticeable how parochial divisions within the scholarly world have thwarted a proper understanding of Collingwood's philosophy of history" (Van der Dussen 2013: 308). The same holds true with respect to *An Essay on Metaphysics*, one could observe, since the reception of this book illustrates that the divide between philosophy and history is not so much reconciled but rather emphasized. One could consider this to be the sad fate of *An Essay on Metaphysics*, since it should indeed be considered an example of the *rapprochement* between philosophy and history as conceived by Collingwood. With *An Essay on Metaphysics* he has anyhow written an exceptional and important book about a subject of great philosophical and historical interest that historians are not accustomed to study.

Notes

1. The quotation is from a section in PH entitled 'Freedom' (PH: 98–103). It was included by Knox in the Epilegomena of IH, with the title 'History and Freedom' (IH: 315–20). The reason that the quotation is taken from PH is that Knox has made two minor changes in the first sentence. On other occasions Knox has interfered with the text more seriously, however. For the first two paragraphs of 'History and Freedom' are of Knox's own making (IH: 315), as is the case with the passage on the bottom of p. 319, from 'I wish to point out' to 'that claim is groundless'. For a more extensive discussion of the way Knox sometimes manipulated the texts of Collingwood's manuscripts, see Jan van der Dussen, 'Collingwood's "Lost" Manuscript of *The Principles of History*', in (Van der Dussen 2016), in particular 43–5, 66–9.
2. In a footnote the original text is given, which is from Spinoza's *Tractatus Theologico-Politicus*, I, iv.: "I have striven not to laugh at human actions, not to weep at them, nor to hate them, but to understand them."

3. In a footnote the original text is given, which is from *Fifteen Sermons Preached at the Rolls Chapel*, N.7,16: "Things and actions are what they are, and the consequences of them will be what they will be. Why then should we desire to be deceived?"
4. By scepticism Knox refers to the fact that in Collingwood's view metaphysical presuppositions are neither true nor false, and by dogmatism that our attitude to our own absolute presuppositions is to be one of 'unquestioning acceptance'.
5. Only the last two essays will be dealt with here.
6. Collingwood declares, referring to the absolute presuppositions of Newtonian, Kantian, and Einsteinian physics, that "every detail in these respective sciences depends on what absolute presuppositions they respectively make. But this does not mean that it depends on these presuppositions' being thought true, or that the truth of the conclusions arrived at depends on the presuppositions' being in fact true. For the logical efficacy of a supposition does not depend on its being true, nor even on its being thought true, but only on its being supposed … … It is a mistake, therefore, to fancy that by investigating the truth of their absolute presuppositions a metaphysician could show that one school of science was fundamentally right and another fundamentally wrong" (EM: 52–3).
7. It reminds one of the way Walter Cronkite used to end the CBS Evening News—that is, recent history—with the phrase "That's the way it is".
8. "[I]t [psychology as pseudo-science] is completely discredited among those (historians, &c.) whose business is to study human thought in its actuality" (PA: 171).
9. That is indeed the case, and Collingwood also referred in IH, in his example of the devil-fearer, to his belief, even going so far as saying of the modern historian who does not believe in devils "that too is only a belief he has accepted" (IH: 318).
10. After the passage where Collingwood says of science that it "seems to be … the simplest and most obvious case in which progress exists and is verifiable", he says: "For this reason, those who have believed most strongly in progress have been much in the habit of appealing to the progress of science as the plainest proof that there is such a thing, and often, too, have based their hope of progress in other fields on the hope of making science the absolute mistress of human life" (IH: 332). Though Collingwood speaks in the past tense, it is the position Martin actually advocates.
11. For an extensive discussion of Collingwood's views on historical process and development, see (Van der Dussen 2012: 45–8, 62–7, 251–62, 331–3; or in 1981 ed.: 53–6, 66–72, 269–81, 353–5).
12. The term 'historicism' has been the subject of much confusion. For besides referring to the German *Historismus*, it may also refer to the notion of

historicism as used by Popper in his *The Poverty of Historicism* and *The Open Society and its Enemies*. For a survey of the confusion concerned—also as regards the terms historicism and historism—see (Van der Dussen 2012: 48–9; or 1981: 377–8).

13. The German text reads: "Man hat der Historie das Amt, die Vergangenheit zu richten, die Mitwelt zum Nutzen zukünftiger Jahre zu belehren, beigemessen: so hoher Aemter unterwindet sich gegenwärtiger Versuch nicht: er will bloss zeigen, wie es eigentlich gewesen" (Von Ranke: 1824, v–vi).
14. The German text reads: "Eine solche gleichsam mediatisirte Generation würde an und für sich eine Bedeutung nicht haben; sie würde nur insofern etwas bedeuten, als sie die Stufe der nachfolgenden Generation wäre, und würde nicht in unmittelbarem Bezug zum Göttlichen stehen. Ich aber behaupte: jede Epoche ist unmittelbar zu Gott, und ihr Wert beruht gar nicht auf dem, was aus ihr hervorgeht, sondern in ihrer Existenz selbst, in ihrem eigenen Selbst" (Von Ranke 1888: 5).

Bibliography

Barker, John. 1982. *The Superhistorians. Makers of Our Past*. New York: Charles Scribner's.

Beaney, Michael. 2005. Collingwood's Conception of Presuppositional Analysis. *Collingwood and British Idealism Studies* 11 (2): 41–114.

Bolingbroke, Lord Viscount. 1752. *Letters on the Study and Use of History*. London: A. Millar.

Breisach, Ernst. 1983. *Historiography. Ancient, Medieval & Modern*. Chicago: University of Chicago Press.

Collingwood, R.G. 1938. *The Principles of Art*. Oxford: Oxford University Press.

———. 1993. *The Idea of History* [1946]. Ed. Jan van der Dussen, Rev. ed. Oxford: Oxford University Press.

———. 1998. *An Essay on Metaphysics* [1940]. Ed. Rex Martin, Rev. ed. Oxford: Oxford University Press.

———. 1999. *The Principles of History and Other Writings in Philosophy of History*. Ed. W.H. Dray and W.J. van der Dussen. Oxford: Oxford University Press.

———. 2005a. *An Essay on Philosophical Method* [1933]. Ed. James Connelly and Giuseppina D'Oro, Rev. ed. Oxford: Oxford University Press.

———. 2005b. *The Philosophy of Enchantment. Studies in Folktale, Cultural Criticism, and Anthropology*. Ed. David Boucher, Wendy James and Philip Smallwood. Oxford: Oxford University Press.

———. 2013. *An Autobiography* [1939] *and Other Writings. With Essays on Collingwood's Life and Work*. Ed. David Boucher and Teresa Smith, Rev. ed. Oxford: Oxford University Press.

Connelly, James. 2013. Collingwood Controversies. In Collingwood, R.G. *An Autobiography and Other Writings. With Essays on Collingwood's Life and Work*, ed. David Boucher and Teresa Smith, Rev. ed., 399–425. Oxford: Oxford University Press.

Connelly, James, Peter Johnson, and Stephen Leach. 2015. *R.G. Collingwood: A Research Companion*. London: Bloomsbury.

Frege, G. 1997. On Sense and Reference ['Über Sinn und Bedeutung']. In *The Frege Reader*, ed. M. Beaney, 151–171. Oxford: Oxford University Press.

Geyl, Pieter. 1974. *Debates with Historians*. Glasgow: Collins Fontana.

Hume, David. 1963. In *Enquiries Concerning the Human Understanding and Concerning the Principles of Morals*, ed. L.A. Selby-Bigge, 2nd ed. Oxford: Oxford University Press.

Krausz, Michael. 1972. The Logic of Absolute Presuppositions. In *Critical Essays on the Philosophy of R.G. Collingwood*, ed. Michael Krausz, 222–240. Oxford: Oxford University Press.

Martin, Rex. 1981. Collingwood's Doctrine of Absolute Presuppositions and the Possibility of Historical Knowledge. In *Substance and Form in History*, ed. L. Pompa and W.H. Dray, 81–106. Edinburgh: Edinburgh University Press.

———. 1995. Collingwood's Claim that Metaphysics Is a Historical Discipline. In *Philosophy, History and Civilization*, ed. David Boucher, James Connelly, and Tariq Modood, 203–245. Cardiff: Cardiff University Press.

———. 1998. Introduction to Collingwood, R.G. *An Essay on Metaphysics*, xv–xcv. Oxford: Oxford University Press.

———. 2013. From Method to Metaphysics. In Collingwood, R.G. *An Autobiography and Other Writings. With Essays on Collingwood's Life and Work*, ed. David Boucher and Teresa Smith, Rev. ed., 353–375. Oxford: Oxford University Press.

Stern, Fritz. 1956. *The Varieties of History. From Voltaire to the Present*. New York: Meridian Books.

Toulmin, Stephen. 1972a. Conceptual Change and the Problem of Relativity. In *Critical Essays on the Philosophy of R.G. Collingwood*, ed. Michael Krausz, 201–221. Oxford: Oxford University Press.

———. 1972b. *Human Understanding*. Princeton: Princeton University Press.

Van der Dussen, Jan. 2012. *History as a Science. The Philosophy of R.G. Collingwood*. Dordrecht: Springer. (1st ed. 1981).

——— 2013. Collingwood's Philosophy of History in the Year of His *An Autobiography*. In Collingwood, R.G. *An Autobiography and Other Writings. With Essays on Collingwood's Life and Work*, ed. David Boucher and Teresa Smith, Rev. ed., 305–333. Oxford: Oxford University Press.

———. 2016. *Studies on Collingwood, History and Civilization*. Cham/Heidelberg: Springer.
Von Ranke, Leopold. 1824. *Geschichten der romanischen und germanischen Völker von 1494 bis 1535*. Leipzig/Berlin: Brokhaus.
———. 1888. *Weltgeschichte* IX, 2nd section, *Über die Epochen der neueren Geschichte*. Ed. A. Dove. Leipzig: Duncker & Humblot.

CHAPTER 5

Presuppositions and the Logic of Question and Answer

Vasso Kindi

Collingwood's logic of question and answer is a direct response to propositional logic which included, in Collingwood's view, not only the symbolic logic of the late nineteenth and early twentieth century but also the traditional, 'idealistic', logic of the eighteenth and nineteenth centuries (A: 34). Collingwood took logic to be the science of thought which he believed comprises three kinds of activities: questioning, supposing, and asserting. He was opposed to taking only propositions, statements, or judgements expressed in indicative sentences as the object of logic and believed that logic is supposed to study not only asserting but all the activities of thought as they are expressed in language (EM: 422). From his perspective, propositions—which he disparaged as the "ghostly double of the grammarian's sentence" (A: 34)—are answers to questions and both questions and answers, which are strictly correlative, ought to be considered as objects of logic. In his view, there is no one-to-one correspondence between propositions and indicative sentences. One sentence does not always express one proposition and indicative sentences in fiction do not express propositions that are true or false. Collingwood also thought that propositions are not

V. Kindi (✉)
National and Kapodistrian University of Athens, Athens, Greece
e-mail: vkindi@phs.uoa.gr

K. Dharamsi et al. (eds.), *Collingwood on Philosophical Methodology*, Philosophers in Depth,
https://doi.org/10.1007/978-3-030-02432-1_5

true or false in isolation, as propositional logic presupposes, but they are right or wrong in response to a question. The unit of thought should not be taken to be the proposition but something bigger and more complex, a constellation of suppositions, questions and answers. The meaning of a proposition does not belong to a proposition "in its own right" (A: 33) but it is relative to the question the proposition answers. In that sense, two propositions that are considered contradictory by propositional logic may not be since they may be answers to two different questions. Only if they answer the same question can they contradict one another (ibid.). Collingwood believed that the requirement of a one-to-one correspondence between propositions and indicative sentences led to the doomed quest for a "well-constructed and well-used language" (A: 35) exemplified by the "typographical jargon of *Principia Mathematica*" (A: 36).

Collingwood names Socrates, Plato, Bacon, Descartes, and Kant as forerunners in the tradition of drawing attention to the significance of questions in philosophy. He points to the Socratic sustained questioning and dialectic and Plato's dialogues where thinking is described as a "dialogue of the soul with itself" (A: 35). He also highlights Bacon's subjection of nature to questioning and even torture so that it properly responds (EM: 238–39), Descartes's systematic doubt, and Kant's effort, as Collingwood saw it, to put metaphysics on the path of scientific progress by raising definite questions "in an orderly and systematic way, instead of arguing blindly" (EM: 239). All these philosophers, Collingwood thought, have said that "knowledge comes only by answering questions, and that these questions must be the right questions and asked in the right order" (A: 25).

One may of course doubt whether all these philosophers were doing the same thing as Collingwood. The dialogue form, for instance, adopted only by Plato in the group of philosophers that Collingwood mentions, may serve several ends that are not within Collingwood's purview. It may be a more lively style of exposition, or of proving a point, in comparison to a treatise,[1] and so more suitable for instruction. Neither of these features are here of interest to Collingwood. In fact, the dialogue form of Plato's work has long drawn the attention of scholars who have pointed to ends other than that which Collingwood points to. Alex Long, for example, distinguishes "three processes which a philosopher like Plato might have thought best served by dialogue: intellectual discoveries, teaching such discoveries to others and converting others to the life of philosophy" (Long 2008: 48). None of these goals feature in Collingwood's work. Besides, according to Richard Kraut (2017) the role played by questions and answers in Plato's dialogues "is never the same from one dialogue to another. (*Symposium*, for example, is a series of speeches, and there are also lengthy speeches in *Apology*, *Menexenus*, *Protagoras*, *Crito*, *Phaedrus*, *Timaeus*, and *Critias*; in fact, one might reasonably question whether these works are properly called dialogues)".

Closer to Collingwood's project with regard to questions and answers are Bacon's and Kant's philosophy. Collingwood said that he was influenced by his work in archaeology to develop his logic of question and answer.

> My work in archaeology, as I have said, impressed upon me the importance of the 'questioning activity' in knowledge [...] The effect of this on my logic was to bring about in my mind a revolt against the current logical theories of the time, a good deal like that revolt against the scholastic logic which was produced in the minds of Bacon and Descartes by reflection on the experience of scientific research, as that was taking new shape in the late sixteenth and early seventeenth centuries. (A: 30)

Both Collingwood and Bacon used their professional experience to propose a change in logic, Bacon, as a jurist and natural historian,[2] with his *Novum Organum*, which was to replace Aristotle's *Organon*,[3] and Collingwood as an archaeologist, with his logic of question and answer which was to replace the standard logic of his time. Both laid emphasis on questions as a means of interrogating nature and acquiring knowledge. Collingwood, however, wanted to stress that we test hypotheses, that is, suppositions that we entertain, and ask questions to get answers. We do not simply rely on observations. Bacon, on the other hand, exalted the importance of sense perceptions[4] and stressed that the various "idols of the mind" and "anticipations of nature", that is, various preconceived ideas, ought to be kept in check and, if possible, removed. Still, because the light of the senses cannot follow nature in her wanderings and penetrate deeply into her labyrinthine intricacies, Bacon suggested the use of mechanical aids to help uncover the secrets of nature. To this end he did not hesitate to recommend questioning, harassing, and vexing nature.[5] Following, among others, Leibniz, Goethe, and Cassirer, who attributed to Bacon the idea of putting nature on the rack (Merchant 2006: 526, 527), Collingwood suggested that "[i]t was Bacon who insisted that the science of nature begins when man begins 'putting nature to the question' (that is, the torture; Bacon was a lawyer and knew what 'the question' meant in his own profession), extorting from her an answer to the questions he chose to ask, instead of contenting himself with noting down whatever she elected to reveal" (EM: 38–39).[6] Bacon uses questions to hunt down nature and learn its secrets, to make her confess what she keeps to herself.

Kant refers to Bacon in his *Critique of Pure Reason* and uses a similar metaphor. He says that reason

> must not allow itself to be kept, as it were, in nature's leading-strings but must itself show the way with principles of judgment based upon fixed laws, constraining nature to give answers to reason's own determining. [...] Accidental observations, made in obedience to no-previously thought out plan, can never be made to yield a necessary law, which alone reason is concerned to discover. Reason, holding in one hand its principles, according to which alone concordant appearances can be admitted as equivalent to laws, and in the other hand the experiment which it has devised in conformity with these principles, must approach nature in order to be taught by it. It must not, however, do so in the character of a pupil who listens to everything that the teacher chooses to say, but of an appointed judge who compels the witnesses to answer questions which he has himself formulated. (Bxiii)

According to Kant, we should not grope around randomly (Bxiv) but demand that nature responds to the particular questions posed. This attitude is in line with Kant's Copernican revolution, which requires that the objects of our senses befit our a priori forms and concepts so that they can be experienced and known and, in that sense, it is closer to Collingwood's than Bacon's account. Collingwood wants nature to conform to the questions/hypotheses that we have while Bacon, who rejects, like Kant, causal experience as groping in the dark (Bacon 2000: 67, 82), treats questioning more as a probing tool. He speaks of the "hunt for nature" (ibid.: 187) and compares his methodology to the 'Hunt of Pan' (cf. Eamon 1994: 286) suggesting that we go after nature to learn what she has to tell us. His questioning is supposed to come up with rules that will allow us to advance knowledge and master nature. Kant, on the other hand, lays emphasis, as Collingwood does, on the presuppositions of knowledge, on what makes knowledge possible.

Collingwood himself uses the metaphor of hunt and prey. In his *An Essay of Metaphysics,* he distinguishes between low- and high-grade thinking (associated with animal and human mind respectively) and says that at the lowest level "one is conscious only of 'intuiting' or 'apprehending' what presents itself to one's mind". At the higher level, however,

> one becomes aware of a mental hunger that is no longer satisfied by what swims into one's mouth. [...] One swims about hunting for it. The ranging of the mind in search of its prey is called asking questions. To ask questions, knowing that you are asking them, is the first stage in high-grade thinking: the first thing that distinguishes the human mind from the sea-anemone mind of the 'realist' theory of knowledge. The second stage is not merely to hunt one's prey but to hunt it cunningly. To hunt it at all is to ask questions; to hunt it cunningly is to ask questions with skill, or scientifically. (EM: 37–38)

Here Collingwood adopts Bacon's terminology of hunting and prey, and it may be said that he also endorses Bacon's quest for a methodology that will help pursue nature more efficiently by his reference to cunning, skill, and scientific procedures. But the emphasis is different in the two accounts. It seems that Bacon would have been satisfied if nature were to swim into his mouth and divulge everything. Collingwood, on the other hand, contrasts his account of asking questions to the way the 'realists' or 'minute philosophers'[7] understand knowledge and the human mind. In his view, 'realist' philosophers in Oxford at the time, such as John Cook Wilson and H. A. Prichard, think of knowing as "a simple intuiting or a simple apprehending of some reality" (A: 25), for fear that if the knowledge of things depends in any way on the mind of the investigators, the process of knowing would become not a process of discovery but a process of constructing. Collingwood thinks that the realist theory of knowledge does not distinguish the human mind from the sea-anemone (EM: 38), which "on its rock receives whatever nourishment the wash of the waves may bring", as William James, who also used the sea-anemone metaphor, put it (James 1996: 64).[8]

In opposition to the realists, Collingwood argued that knowledge of nature depends on the particular questions posed. He gives the example of archaeologists who "never dug a trench without knowing exactly what information they were looking for; they knew both that this information was the next thing they needed for the progress of their study, and also that this trench would give it to them" (A: 124). He calls these methods Baconian, interpreting the Baconian revolution as a revolution that "converts a blind and random study into one where definite questions are asked and definite answers insisted upon" (ibid.). Yet, although Bacon thought that we need "a thread to guide our steps" (2000: 10), a thread provided by his method, he insisted that "we stay faithfully and constantly with things, and abstract our minds no further from them than is necessary for the images and rays of things to come into focus" (ibid.: 11). Bacon distrusted the mind because of its mental habits and its spontaneous leaps to abstractions and general notions and wanted to leave little "to the power and excellence of the intelligence" (ibid.). In contradistinction to Kant, who did not want to approach nature as a pupil who expects to learn from nature, Bacon recommended "humility in discovery" (ibid.) and thought that "nature is conquered only by obedience" (ibid.: 24, 33; cf. 100).

It seems, then, that Collingwood's emphasis on questions is more similar to Kant's concern with presuppositions than to Bacon's interest in the secrets of nature.[9] Collingwood's affinity to Bacon lies more in the attention they both give, on the one hand, to particulars as opposed to abstractions and general notions and, on the other, on the development of a method that would make the study of nature scientific. Bacon concentrated on devising and refining his inductive method while Collingwood associated scientific thinking with the orderly sorting out of muddles and perplexities.

> In unscientific thinking our thoughts are coagulated into knots and tangles; we fish up a thought out of our minds like an anchor foul of its own cable, hanging upside-down and draped in seaweed with shellfish sticking to it, and dump the whole thing on deck quite pleased with ourselves for having got it up at all. Thinking scientifically means disentangling all this mess, [...] (EM: 22)[10]

Both Bacon and Collingwood were also averse to hasty generalizations. Bacon rejected abstract generalities and notions formed by hearsay and public opinion (Bacon 2000: 36, 37).[11] Collingwood said that his inclinations always led him "towards detail than towards generalization" (A: 74). The untangling of the knots of thought to a series of specific questions and answers and the ensuing renunciation of propositional logic, which deals with propositions that are atemporal abstractions, serve the same purpose, namely, to direct attention to the particular, the determinate, and the concrete.

* * *

The determination and concreteness that interests Collingwood is, each time, established by the particular context in which a question or an answer is embedded, a context that should not be understood, however, as a mere supplement of, or an addition to, supposedly eternal propositions. Collingwood rejected propositions as abstract entities and thought that there are no eternal philosophical problems (EM: 72; A: 60). He also thought that there are no universals that may be accounted for or instantiated differently. For instance, Plato's state, that is, the ancient *polis*, and Hobbes's state, that is, the absolutist state of the seventeenth century, are not two different accounts of the same idea. They are two different things that are related between them by a historical process which has turned one into the other (A: 61–62). Context, for Collingwood, is not an already given, separate, background structure that is brought to bear on propositions that already have their own intrinsic meaning. The propositions that are each time under consideration are already imbued with meaning by their correlative presuppositions that make them possible. This is what makes them always concrete. As Collingwood put it, the thoughts that are connected to a thought expressed, "are not merely its context, they are its presuppositions" (EM: 21). In Collingwood's view,

> [m]eaning, agreement and contradiction, truth and falsehood, none of these belonged to propositions in their own right, propositions by themselves; they belonged only to propositions as the answers to questions: each proposition answering a question strictly correlative to itself. (A: 33)[12]

The metaphysicians, according to Collingwood, are supposed to highlight the presuppositions that "are doing their work in darkness" (EM: 43) and to bring them forward to consciousness. But they are not thereby supposed to offer us new knowledge. In a manner that is reminiscent of Plato,[13] Collingwood vouches, as a metaphysician, to remind us of what we already know: "In expounding these propositions I shall not be trying to convince the reader of anything, but only to remind him of what he already knows perfectly well" (EM: 23). Here, Collingwood resembles not just Plato but also Wittgenstein who said that "[t]he work of the philosopher consists in marshaling reminders for a particular purpose" (Wittgenstein 2009: §127).[14] Wittgenstein also said that he was not trying to persuade his interlocutors to change their opinions (Wittgenstein 1975: 103). He had no opinion on the questions he discussed; he was just recommending a method and was not trying to teach any new truths (Wittgenstein 2001: 97).

Michael Beaney, in 'Collingwood's Conception of Presuppositional Analysis' (2005), criticizes Collingwood for tensions and inconsistencies in his account of absolute presuppositions. More particularly, he finds fault with Collingwood's view that absolute presuppositions are not propounded and are not propositions. Beaney says that they can indeed be stated and questioned and so, *pace* Collingwood, they are propositions just like the relative presuppositions. He also argues that, since, according to Collingwood, metaphysical sentences, which are historical propositions and include in their expression the absolute presuppositions, are true or false, they need to be already meaningful, which implies that the part which expresses the absolute presupposition is also meaningful and so a proposition. For instance, the metaphysical sentence "Kant absolutely presupposed that all events have causes" is true and, therefore, meaningful, which means, according to Beaney, that 'all events have causes', is also meaningful. This conclusion clashes with Collingwood's thesis that absolute presuppositions lack truth value, which should imply, according to Beaney, that they are meaningless.

Beaney sees a contradiction, but I see a misinterpretation of what Collingwood does. Collingwood does not say that we cannot grammatically turn a sentence expressing an absolute presupposition into a question. Of course, we can. But that is not the point of what Collingwood does. Collingwood notes that "[i]t is no answer to the statement 'that question does not arise' if someone says 'it *does* arise because I ask it'" (EM: 394). When Collingwood says, as regards absolute presuppositions, that 'a question does not arise' or that it is meaningless (EM: 393) or nonsense (EM: 26; 408), he does not mean that we are dealing with gibberish, incomprehensible sounds or indecipherable scribbles. He means that questioning absolute presuppositions betrays a misunderstanding of their role in thought (EM: 394, 408). How do we know that a presupposition is relative, and can be questioned, or absolute and cannot? Collingwood says that we detect and uncover presuppositions, in general, by interpreting evidence and doing historical research. It is the work of the metaphysician to report what people suppose in a particular historical phase. This kind of work is historical. As for the absolute presuppositions, in particular, Collingwood gives some clues or symptoms. He says that "when an absolute presupposition is touched, the invitation [to justify or abandon it] will be rejected, even with a certain degree of violence" (EM: 44). He also says that "people are apt to be ticklish" when they are asked about their absolute presuppositions (EM: 31). The reason they are ticklish is that it seems inconceivable to them to think without the presuppositions questioned (EM: 413).

In Collingwood's logical universe propositions are different from suppositions (EM: 422–23). Beaney does not distinguish between propositions and suppositions and treats the absolute presuppositions as ordinary propositions, that is, as abstract entities, that have meaning independently of the context in which they occur and as capable of participating in truth-functional substitutions. That is, he treats propositions exactly as Collingwood did not want to treat them. As already stated, according to Collingwood, propositions do not have intrinsic meaning in isolation from the network in which they are embedded. What is more, absolute presuppositions do not circulate in language, they are not actually used; they are taken for granted by those who presuppose them even without realizing that they do (EM: 41, 43, 102). Absolute presuppositions are not themselves propositions. When they are reported in empirical statements, they do not any more function as absolute presuppositions. The empirical propositions are part of the metaphysician's framework, not part of the framework that is made possible by the corresponding absolute presuppositions. Thomas Kuhn, making use of Ian Hacking's idea that different styles of reasoning have different candidates for truth or falsehood (Hacking 2002), says that "[a] statement may be a candidate for truth/falsity with one lexicon without having that status in the others" (Kuhn 2000: 100). Analogously we can say that the absolute presupposition which Beaney treats as a statement may be a candidate for truth/falsity in the metaphysician's framework where it features in empirical propositions, but not in the framework in which it serves as an absolute presupposition.

Beaney ignores Collingwood's claim that "it is nonsense to say, as some modern logicians do say, that supposing is one of various 'attitudes' which we can take up towards a proposition, where a proposition means something which can be either true or false" (EM: 33). Beaney (2005: 73–74) shares the view of these "modern logicians", that is, the logical positivists, and argues that, since absolute presuppositions are the content of propositional attitudes, they are indeed propositions.[15] Collingwood rejects this as "merely a device for imposing on unwary readers the dogma ... that all presuppositions are relative, or that there are no absolute presuppositions" (EM: 33). Beaney (2005: 85) thinks that, to identify the absolute presuppositions of a framework, one must understand the propositions that express them, which in turn involves passing judgement on them as to whether they are true or false. This is a Davidsonian idea which says that in order to understand a different language or conceptual scheme, you need to presuppose the truth and rationality of most of the beliefs held by the speakers of that language.

> If an interpretation or analysis results in their apparently holding a lot of patently false beliefs, then the principle of charity suggests that the interpretation or analysis must be revised to make better sense of their beliefs. This is a widely accepted principle in contemporary philosophy of language, familiar, in particular, from the writings of Donald Davidson: interpret so as to maximise the truth and rationality of the subject's beliefs. (Beaney 2005: 85–86)

But why should one accept Davidson's view? First of all, one may want to distinguish between understanding and reporting in the way Thomas Kuhn (1982) distinguished between interpreting or learning a language on the one hand and translating on the other. Collingwood himself distinguishes between comprehending and passing judgement, although he says that the two attitudes are inseparable:

> Comprehension and criticism, or understanding what the writer means and asking whether it is true, are distinct attitudes, but not separable [...] one is prior to the other: the question whether a man's views are true or false does not arise until we have found out what they are. (EPM: 217)

Secondly, practitioners, such as anthropologists or historians, who actually engage in understanding, interpreting, and translating radically different languages do not at all accept Davidson's requirement of charity which enjoins that we presuppose extensive similarities with the language to be interpreted.[16] In fact, they recommend the exact opposite. For instance, the historian Herbert Butterfield maintained that

> [i]t is not for him [the historian] to stress and magnify the similarities between one age and another ... Rather it is his work to destroy those very analogies which we imagined to exist. (Butterfield 1965: 10)

Similarly, Thomas Kuhn has said that

> the past of science should be approached as an alien culture, one that the historian strives first to enter and then to make accessible to others. Entry into another culture, scientific or not, is regularly resisted ... and the standard form of resistance is to carry one's own culture with one and assume that the world conforms. (Kuhn 1984: 250)[17]

Collingwood criticizes even the practitioners who focus on similarities. For instance, he is critical of the assumption, made by the anthropologist James G. Frazer, that similarities between customs "indicate substantial identities" (PE: 152) and argues that perceived similarities may be superficial and misleading. He says that there is a tendency in certain schools of anthropology "to slur over details, emphasize similarities at the expense of diversities, and thus reduce to a spurious uniformity things which may well be essentially different" (PE: 181). Collingwood's emphasis is always on the particular and the concrete.

* * *

As stated above, Collingwood distinguishes between absolute presuppositions that ordinary people and scientists make, and which can only catachrestically be called propositions since they are not true or false, and the propositions that a metaphysician uses to report the absolute presuppositions of a particular period. The latter are historical propositions and have the form "In such and such a phase of scientific thought it is (or was) absolutely presupposed that ..." (EM: 55). This prefix, which is supposed to always introduce the statement of absolute presuppositions, is called by Collingwood "the metaphysical rubric" and can be omitted for reasons of convenience in case it is assumed that the readers know that, even without it, the metaphysicians state historical facts and do not themselves make these absolute presuppositions. The metaphysicians are not supposed to ask whether the absolute presuppositions are true or false. They are only supposed to offer an exposition, a presentation (EM: 383). When these historical propositions are true, they assert that the people of the period referred to did indeed have these absolute presuppositions; when they are false they state that the people referred to did not actually make these presuppositions (EM: 411–12). The metaphysician has erred. Truth and falsity characterize the metaphysicians' historical propositions and not the absolute presuppositions themselves.[18]

Metaphysicians, being historians, realize that the constellations of absolute presuppositions differ significantly across historical periods and cultures (EM: 86n1, 379, 419). They can compare them to identify similarities and differences[19] and, thereby, overcome "the parish pump idea that the metaphysical problems of his own generation or, more likely, the one next before his own are the problems that all metaphysicians have been worrying ever since the world began" (EM: 72), but they do not really have a real standard of comparison between them (EM: 391). They cannot prove or proclaim which one is better or superior. So, the question 'which set of absolute presuppositions is true or better?' is, according to Collingwood, meaningless (EM: 46, 392–93). It is meaningless for the metaphysician, but it is also meaningless for the people who make these presuppositions because they are usually not even aware that they have them. They do not adopt them because they have compared them with others and have been convinced of their truth and superiority. Actually, absolute presuppositions are not entertained to be chosen (EM: 27, 31, 48, 392, 415).[20] They cannot be resisted; they are forced on us the way the redness of red things is forced on us in immediate experience (EM: 415). One can hardly conceive how it would be possible to think without them (EM: 413). Absolute presuppositions are inculcated by time-honoured institutions in a long and slow process which includes education (EM: 197, 413). They are neither derived nor judged by experience (EM: 193, 197) and they are not argued for (EM: 173). So,

> it is impossible to decide which of two opposing systems of metaphysical principles shall be regarded as true, in the sense of holding them up before one and deciding out of an open mind which one shall presuppose, because no one is ever in the situation of open-mindedness. (EM: 414)

Collingwood's account of absolute presuppositions gave rise to the following concern: How do absolute presuppositions change? Given that they are not contemplated to be justified and chosen by the agents who hold them, and that they cannot be evaluated by metaphysicians as to their truth or falsity, it seems that their change cannot be rationally accounted for. As Stephen Toulmin (1972: 210–11) put it, if we cannot justify the transitions rationally, that is, by appealing to reasons and arguments, then, "all we have left to do is to explain them causally", that is by appealing to forces and compulsions. Collingwood, as Toulmin acknowledged (ibid.: 210), would reject both alternatives. It has already been said that for Collingwood, adopting a set of presuppositions is not a matter of rational decision. But Collingwood was equally opposed to studying the change of absolute presuppositions causally. He may have had a broad concept of 'cause',[21] but he did not want to assign the subject matter of metaphysics to natural science. This is the reason he rejected the suggestion to relegate the study of absolute presuppositions to psychology which was presented to him as the science of how we think, not only consciously but also unconsciously. The unconscious part is particularly important because Collingwood, replying to an inquiring friend (namely, A.D. Lindsay, Master of Balliol College, Oxford; EM: 350–55) said that a change or modification of absolute presuppositions is not "consciously devised but created by a process of unconscious thought" (EM: 48). And yet, Collingwood dismissed the proposal to leave the matter to psychology despite the fact that he considered the information that "modern analytic methods have enabled it [psychology] to probe into the deepest recesses of the unconscious, and to reveal the existence and the dynamic power of thought-processes which, because the people in whom they occur are wholly unaware of them, have never before been studied" (EM: 102). Collingwood dismissed psychology as a science of thought[22]—he called it pseudo-science—because, in his view, it treats thought not as a self-critical activity but as reducible to sensations, feelings, and appetites. Collingwood wants to preserve the normative dimension of thought, that is, the distinction between thinking truly and thinking falsely (EM: 117) and thinks that logic and metaphysics are the appropriate disciplines to study it.

But how does metaphysics study the change of absolute presuppositions? On the one hand Collingwood said that metaphysics carries out a historical inquiry, and a historical inquiry does not account for historical events in terms of physiological, psychological, climatic, geographical, and, in general, natural causes (EM: 98). The causes of historical events, he said, should not be sought "outside history" (EM: 73) but rather within history, that is, in human activity and thought. On the other hand, if people are not ordinarily aware of their absolute presuppositions and they do not choose to change them (EM: 48), how can the history of absolute presuppositions be the history of thought as Collingwood required of any historical inquiry? The situation is as follows: We cannot account for historical events in terms of natural causes and we cannot account for absolute presuppositions in terms of conscious thought. How to break the impasse? One way is to say that metaphysics is not really history but rather conceptual analysis, as Rex Martin has argued (1998: li-lii). But, although metaphysics involves logical analysis, it is not confined to it. Absolute presuppositions come in constellations and the relations between the various items in these constellations are not logical relations (EM: 67). If they were, we would not be dealing with absolute presuppositions but with relative ones. As Collingwood says,

> [t]he presence of a given item in a metaphysical system is a question of fact, not a question of logic. The question *what* relations there are between the items of a metaphysical system is again for the metaphysician a question of fact. He has to settle it not on any abstract logical or dialectical methods of his own devising, but by studying the actual way in which the people whose thought he is analyzing treat their presuppositions. (EM: 384, emphasis in the original)

Metaphysics is not a "deductive science" (EM: 65–67, 76). Principles or presuppositions that, in the abstract, may appear contradictory to a metaphysician, may not, in actual practice, be in conflict at all, if they apply to different domains (EM: 385, 424). And this can only be found out empirically, that is, historically. The other alternative to break the impasse would be to say that human activity, which falls within history, need not be individual, deliberate, and conscious. It can be habitual, collective, and impersonal (IH: 219).[23] This is the line that Collingwood takes. In his view, acquiring absolute presuppositions is not a matter of compulsion, as it would be the case if we were dealing with natural causes, but a matter of inculcation and habit which are again human activities. The change of absolute presuppositions can be studied historically, which for Collingwood means, studied as a manifestation of historical dynamics. Because the constellation of absolute presuppositions is not a logically coherent structure, it is always in unstable equilibrium with a lot of internal strains and, therefore, with the seeds of change in itself (EM: 74).[24] The various elements are kept together with great effort and compromise. Collingwood suggests a "reformed metaphysics" which

> will conceive any given constellation of absolute propositions as having in its structure not the simplicity and calm that characterize the subject-matter of mathematics but the intricacy and restlessness that characterize the subject-matter, say, of legal or constitutional history. (EM: 77)

So, Collingwood's proposal of reform in philosophy, through the logic of question and answer and through his treatment of presuppositions and absolute presuppositions, aims at recommending a situated and historical approach which stands opposed to the reduction of philosophy to natural science on the one hand and its sublimation to the heights of abstraction, universalism, and a prioricity on the other. Both natural science and a priori philosophy lose sight of the particular and the concrete and resort too easily to sweeping generalizations and homogenizing abstractions.

Notes

1. See (Hume 2007: 4): "Any question of philosophy, on the other hand, which is so *obscure* and *uncertain*, that human reason can reach no fixed determination with regard to it; if it should be treated at all; seems to lead us naturally into the style of dialogue and conversation."
2. Harvey Wheeler has argued that "Bacon's science derived ultimately from his jurisprudence", cited in (Pesic 1999: 91).
3. Bacon calls his art "Interpretation of Nature" and says that it is an "art of logic" (2000: 15) which is to replace the ordinary one. Those who use the ordinary logic, says Bacon, "defeat and conquer their adversary by disputation; we conquer nature by work" (ibid.: 16).
4. "[W]e have made the senses (from which, if we prefer not to be insane we must derive everything in natural things) sacred high priests of nature and skilled interpreters of its oracles" (Bacon 2000: 18).
5. "[W]e are making a history not only of nature free and unconstrained (when nature goes its own way and does its own work), such as a history of the bodies of heaven and the sky, of land and sea, of minerals, plants and animals; but much more of nature confined and harassed, when it is forced from its own condition by art and human agency [...] nature reveals herself more through the harassment of art than in her own proper freedom" (Bacon 2000: 20–21). Cf. "the secrets of nature reveal themselves better through harassments applied by the arts than when they go on in their own way" (ibid.: 81).
6. *Pace* Collingwood but also some feminist scholars, Peter Pesic (1999: 84) insists that "[t]he interrogation [of nature] requires handcuffs and chains, but it is not a scene of torture." Bacon compares nature to the mythical figure of Proteus who has to be constrained and handcuffed to reveal all that he knows. Pesic (ibid.: 86) cites Bacon who says "if anyone gets annoyed because I call the arts the bonds of nature when they ought rather to be considered its liberators and champions in that in some cases they allow nature to achieve its ends by reducing obstacles to order, then I reply that I do not much care for such fancy ideas and pretty words; I intend and mean only that nature, like Proteus, is forced by art to do what would not have been done without it: and it does not matter whether you call this forcing and enchaining, or assisting and perfecting." Feminist scholars such as Carolyn Merchant, Sandra Harding, and Evelyn Fox Keller have argued that Bacon uses rape metaphors and sexual imagery to speak about knowledge of nature. A characteristic passage from Bacon cited by all these feminist scholars is the following: "For you have but to follow and as it were hound nature in her wanderings, and you will be able when you like to lead and drive her afterwards to the same place again. [...] Neither ought a man to make scruple of entering and penetrating into these holes and corners, when the inquisition of truth is his sole object—as your Majesty has shown in your own example" (Merchant 1980: 168).

7. Chapter III of Collingwood's *Autobiography* is entitled "Minute philosophers" and it discusses the Oxbridge 'realists'. As Michael Beaney (2013: 249) notes, the term 'minute philosophers' was previously used by Cicero and by George Berkeley in his book, *Alciphron, or the Minute Philosopher*. In Beaney's view (ibid.: 260), Collingwood's logic of question and answer has roots in the tradition it was supposed to attack, that is, Oxford and Cambridge realism. Yet, he also recognizes that Collingwood was not just interested in identifying the subject and predicate of a statement through the question that the statement answers, as Cook Wilson used to do in *Statement and Inference* but was also concerned with "the very meaning and truth-value of the statement" (ibid.: 261).
8. James's emphasis is on concepts which help us "go in quest of the absent, meet the remote, actively turn this way or that, bend our experience, and make it tell us wither it is bound. We change its order, run it backwards, bring far bits together and separate near bits, jump about over its surface instead of plowing through its continuity, string its items on as many ideal diagrams as our mind can frame" (James 1996: 64). The Jamesian phraseology regarding experience brings to mind Bacon's language when he talks about how nature needs to be harassed in order to be known.
9. Cf.: "The fundamental presuppositions of modern natural science, or what I am calling its metaphysical presuppositions, are what Kant calls the principles of the pure understanding, e.g., that every event has a cause" (EM: 400). In his *Idea of History* (IH: 237), Collingwood compares wringing answers from nature the way Bacon did it, to putting the historian's authorities in the witness-box. The metaphor of the witness-box that Collingwood uses implies that, in his view, Bacon and Kant, who also use the same metaphor, may be taken to engage in the same kind of investigation. Yet, as I have argued, Kant and Bacon may both emphasize the questioning of nature, but their attitude towards it is significantly different.
10. Cf.: "Philosophy is like unraveling a ball of wool." (Wittgenstein 1975: 220); "Philosophy unravels the knots in our thinking" (Wittgenstein 2005: 311).
11. "The human understanding is carried away to abstractions by its own nature, and pretends that things which are in flux are unchanging. But it is better to dissect nature than to abstract" (Bacon 2000: 45). Cf.: "But our logic instructs the understanding and trains it, not (as common logic does) to grope and clutch at abstracts with feeble mental tendrils, but to dissect nature truly, and to discover the powers and actions of bodies and their laws limned in matter" (ibid.: 219–220).
12. Cf.: "whether a proposition is T or F, significant or meaningless depends on what question it was meant to answer" (A: 39).

13. Recollection (ἀνάμνησις) of what we already know was for Plato the means to learning (μάθησις) and was associated—more explicitly by Aristotle—to thinking and discursive reasoning. It was an active search of what it is to be recalled and was, thus, distinguished from the passive state of memory (μνήμη) which was associated with perception (see Bloch 2007: 72–76). It is noteworthy that Collingwood, in his *An Essay on Philosophical Method*, referring to Socrates, contrasts knowledge from without via perception (by observing facts) and knowledge from within via thinking. According to Socrates's theory, "knowledge was to be sought within the mind, and brought to birth by a process of questioning" (EPM: 10–11).
14. Cf.: "the real difficulty in philosophy is a matter of memory—memory of a peculiar sort" (Wittgenstein 1975: 44). For a comparison of Collingwood and Wittgenstein as regards the description of facts see Kindi (2016).
15. In fairness to Beaney it must be said that he points to ambiguities in the use of the word 'proposition' by Collingwood, one of them being the sense adopted by the logical positivists. But at times it seems that Beaney himself is willing to embrace the same, widely held, understanding of propositions.
16. Cf.: Forster (1998). In opposition to Davidson, Michael Forster argues that important anthropologists and classicists, such as E. Fränkel and Evans-Pritchard, who had extensive experience in interpreting languages, behaviour, and texts, "firmly reject" massive commonality and agreement between remote cultures and suggests "to turn the tables on Davidson's a priorist argument from his theory of meaning, understanding, and interpretation to the impossibility of discovering radical differences in beliefs and concepts —to accept the discovery of such differences as an established fact and to see in the incompatibility of this fact with Davidson' s theory a further argument against the *theory*" (Forster 1998: 151–152, emphasis in the original).
17. I owe Butterfield's and Kuhn's quotes to Theodore Arabatzis who made use of them in a talk on Davidson's principle of charity and on how differently this principle is perceived by philosophers on the one hand and historians on the other.
18. The logical positivists also thought that absolute presuppositions, such as 'All events have causes" or "God exists", lack truth value because they are unverifiable. But, *pace* Collingwood, they thought that they are nonsensical pseudo-propositions. According to Collingwood, however, if we were to think, as positivists do, that absolute presuppositions are nonsensical, then these pseudo-propositions could not, of course, serve as presuppositions and thinking would be left confused since systematic or scientific thinking depends on presuppositions and requires the disentanglement of a series of questions and answers (EM: 170–71, 233).

19. It has to be stressed that the identification of similarities and differences is not meant by Collingwood to encourage the natural scientist's work of classifications and generalizations.
20. Collingwood distinguishes between assumptions and suppositions. The assumptions, as in the case of mathematics where we begin by making a certain hypothesis, such as 'Let ABC be a triangle', are freely chosen whereas suppositions proper aren't (EM: 27–28, 414–15).
21. Collingwood says that the concept of cause is ambiguous. One of its senses is the one that Collingwood calls 'the historical sense', in which the cause is associated with human activity, that is, deliberate actions and motives (EM: 285–86; IH: 214–15; PH: 191).
22. It should be noted that Collingwood does not dismiss psychology *tout court*. He criticizes Freud—he mentions Freud's *Totem and Taboo*—and other psychologists (EM: 118, 122–132), but he insists that he values the achievements of psychology as a science of feeling (EM: 141).
23. Cf. what Collingwood says about character in history: that changes of character are brought about not only by the deliberate actions of the agent but also by the agent's past, by what has happened to the agent. The agent's past that constitutes the agent's character can be studied historically as any other (PH: 192).
24. The Collingwoodian thesis is reminiscent of Kuhn's idea about the *necessity* of scientific revolutions (Kuhn 1970: 92–110). According to Kuhn, deep change is bound to occur in science because the practice of normal science carries within it the seed of change, namely, the prospect of anomalies which are bound to occur given the expansive nature of normal science on the one hand and its limited resources on the other.

Bibliography

Bacon, F. 2000. *The New Organon*, ed. Lisa Jardine and Michael Silverthorne. Cambridge: Cambridge University Press.

Beaney, M. 2005. Collingwood's Conception of Presuppositional Analysis. *Collingwood and British Idealism Studies* 11 (2): 41–114.

———. 2013. Collingwood's Critique of Oxbridge Realism. In *R. G. Collingwood's An Autobiography and Other Writings*, ed. David Boucher and Teresa Smith, 247–269. Oxford: Oxford University Press.

Bloch, D. 2007. *Aristotle on Memory and Recollection*. Leiden: Brill.

Butterfield, H. 1965. *The Whig Interpretation of History*. New York: W.W. Norton & Company.

Collingwood, R.G. 1982. *An Autobiography* [1938]. Oxford: Oxford University Press.

———. 1993. *The Idea of History* [1946]. Oxford: Oxford University Press.

———. 1998. *An Essay on Metaphysics* [1940], ed. R. Martin. Oxford: Oxford University Press.

———. 1999. *The Principles of History*. Oxford: Oxford University Press.

———. 2005a. *An Essay on Philosophical Method* [1933], eds. G. D'Oro and J. Connelly. Oxford: Oxford University Press.

———. 2005b. *The Philosophy of Enchantment*, eds. D. Boucher, W. James and P. Smallwood. Oxford: Oxford University Press.

Eamon, W. 1994. *Science and the Secrets of Nature. Books of Secrets in Medieval and Early Modern Culture*. Princeton: Princeton University Press.

Forster, M.N. 1998. On the Very Idea of Denying the Existence of Radically Different Conceptual Schemes. *Inquiry* 41: 133–185.

Hacking, I. 2002. Language, Truth and Reason [1982] in His *Historical Ontology*. Cambridge, MA: Harvard University Press, 159–177.

Hume, D. 2007. *Dialogues Concerning Natural Religion* [1779], ed. Dorothy Coleman. Cambridge: Cambridge University Press.

James, W. 1996. *Some Problems of Philosophy* [1911]. Lincoln: University of Nebraska Press.

Kindi, V. 2016. Collingwood, Wittgenstein, Strawson: Philosophy and Description. *Collingwood and British Idealism Studies* 22 (1): 19–43.

Kraut, R. 2017. Plato. In *The Stanford Encyclopedia of Philosophy*, ed. Edward N. Zalta, URL: https://plato.stanford.edu/archives/fall2017/entries/plato/. Accessed 5 Apr 2018.

Kuhn, T.S. 1970. *The Structure of Scientific Revolutions* [1962]. Chicago: The University of Chicago Press.

———. 1982. Commensurability, Comparability, Communicability. *Philosophy of Science Association* 2: 1–19.

———. 1984. Revisiting Planck. *Historical Studies in the Physical Sciences* 14 (2): 231–252.

———. 2000. The Road Since *Structure* [1991]. In *The Road Since Structure*, ed. James Conant and John Haugeland, 90–104. Chicago: The University of Chicago Press.

Long, A. 2008. Plato's Dialogues and a Common Rationale for Dialogue Form. In *The End of Dialogue in Antiquity*, ed. Simon Goldhill, 45–59. Cambridge: Cambridge University Press.

Martin, R. 1998. Editor's Introduction. In *An Essay on Metaphysics*, ed. R.G. Collingwood, xv–xcv. Oxford: Oxford University Press.

Merchant, C. 1980. *The Death of Nature: Women, Ecology, and the Scientific Revolution*. San Francisco: HarperCollins.

———. 2006. The Scientific Revolution and *The Death of Nature*. *Isis* 97: 513–533.

Pesic, P. 1999. Wrestling with Proteus: Francis Bacon and the "Torture" of Nature. *Isis* 90 (1): 81–94.

Toulmin, S. 1972. Conceptual Change and the Problem of Relativity. In *Critical Essays on the Philosophy of R.G. Collingwood*, ed. Michael Krausz, 201–222. Oxford: Clarendon Press.

Wittgenstein, L. 1975. *Wittgenstein's Lectures on the Foundations of Mathematics. Cambridge 1939*, ed. Cora Diamond. Chicago: The University of Chicago Press.

———. 2001. *Wittgenstein's Lectures. Cambridge 1932–1935*, ed. Alice Ambrose. New York: Prometheus Books.

———. 2005. *The Big Typescript. TS 213*, eds. and Trans. C. Grant Luckhardt and Maximilian A. E. Aue. Oxford: Blackwell.

———. 2009. *Philosophical Investigations* [1953], Revised 4th Edition, P.M.S. Hacker and J. Schulte (eds.), G.E.M. Anscombe, P.M.S. Hacker and J. Schulte (trans.). Oxford: Blackwell.

CHAPTER 6

Collingwood, Pragmatism, and Philosophy of Science

Elena Popa

1 Introduction: Collingwood's Metaphilosophy

This chapter places Collingwood's philosophical methodology in the context of contemporary pragmatist approaches to philosophy of science. I argue that Collingwood's method of investigating absolute presuppositions has important similarities to pragmatist contributions to two important topics in the philosophy of science, namely, the debate between realism and anti-realism, and the problem of causation. In this section, I discuss Collingwood's method of investigating absolute presuppositions with focus on two uses—his argument against realism and his analysis of causation. In Sect. 2, I draw a broad connection between Collingwood and pragmatism with emphasis on recent strands of pragmatism that focus on the critique of realism and the primacy of practice. In Sect. 3, I argue that there are significant similarities between Collingwood's critique of realism alongside his broader view on metaphysics and Putnam's critique of externalism. In Sect. 4, I explore the similarities between Collingwood's account of causation and Price's pragmatist approach to causation and

E. Popa (✉)
Asian University for Women, Chittagong, Bangladesh
e-mail: elena.popa@auw.edu.bd

K. Dharamsi et al. (eds.), *Collingwood on Philosophical Methodology*,
Philosophers in Depth,
https://doi.org/10.1007/978-3-030-02432-1_6

critique of causal realism. Section 5 investigates the extent to which Collingwood's work can be classified as pragmatist in the light of the comparisons to Putnam and Price. I conclude by placing the affinities between Collingwood's method and pragmatism in a broader perspective, showing how Collingwood's strategy of investigating philosophical problems matches current ways of approaching issues in the philosophy of science. Insofar as these are important approaches to issues such as causation, or realism, Collingwood's work proves to be relevant to these contemporary debates.

Collingwood's views on the issues of realism and causation belong to his broader metaphilosophical approach, especially with respect to metaphysics. In contemporary debates within the philosophy of science, these topics are also subsumed under metaphysics, although the methods of current metaphysics of science are distinct from Collingwood's proposed account. As I will be pointing out, although the discipline is structured differently, there are significant similarities to Collingwood's method when investigating pragmatist approaches to realism and causation.

Regarding methodology, a notable claim from Collingwood's *Essay on Philosophical Method* holds that philosophy "does not, like exact or empirical science, bring us to know things of which we were simply ignorant, but brings us to know in a different way things which we already knew in some way" (EPM: 161). This passage presents the goal of philosophy as uncovering hidden assumptions, and thus bringing previous knowledge into a different light. I take Collingwood's considerations on metaphysics to fall within this broader scope of his philosophy. While it is subject to debate whether Collingwood changed views between *An Essay on Philosophical Method* and *An Essay on Metaphysics*, I follow interpretations that establish continuity between the two (see Martin 1998: liii–liv).

Concerning metaphysics, in the beginning of *An Essay on Metaphysics*, Collingwood rejects the definition of metaphysics as a science of pure being and proposes a view on metaphysics as "a science of absolute presuppositions" (EM: 41). Collingwood describes propositions as answers to questions; propositions can be propounded, and they can be true or false (EM: 23–25). Collingwood further holds that "every question involves a presupposition" (EM: 25). One of the defining features of presuppositions is that they do not need to be true in order to function (EM: 28). In distinguishing between relative and absolute presuppositions, Collingwood holds that absolute presuppositions are not propositions, since they do not have a truth value (EM: 32). Absolute presuppositions constitute the subject of metaphysics: "the scientist's business is not to propound them, but only to presuppose them. The metaphysician's business [...] is not to propound them, but to propound the proposition that this or that one of them is presupposed" (EM: 33). While there are several aspects of interest concerning Collingwood's proposed reform of metaphysics, for my purposes here, the issue of heuristics and modes of inquiry is central. As D'Oro and Connelly state, "what makes the presupposition absolute is that it cannot be forsaken without at the same time undermining the form of enquiry which it makes possible" (D'Oro and Connelly 2015: 3.2). Martin also emphasizes the role of absolute presuppositions in understanding: "absolute presuppositions constitute the nature of our understanding within a particular universe of discourse and they govern the inquiry we undertake and the explanations we give there in individual cases" (Martin 1998: xxvi–xxvii). This overall view on Collingwood's method and his considerations on the status of metaphysics enable a closer look at the topics stated above: realism and causation.

Concerning realism, Collingwood's proposed reform of metaphysics involves the denial of presuppositionless knowledge. As pointed out by D'Oro: "Collingwood's claim to be practising metaphysics without ontology [...] should be read as putting forward the rather strong claim that there can be no presuppositionless inquiry into reality rather than as stating the rather weaker claim that descriptive metaphysicians should limit themselves to an examination of our ways of thinking and speaking about the world" (D'Oro 2002: 22–23). As a continuation of this project, Collingwood's denial of realism is not what could be labelled as an ontological claim, but rather an epistemological one: "the 'realism' which constitutes the target of Collingwood's critique is not the ontological thesis that there exist mind-independent objects, but the epistemological thesis that there is such a thing as presuppositionless knowledge of reality" (D'Oro and Connelly 2015: 3.2). Collingwood's rejection of realism is explicit in *An Autobiography*:

> For if you know that no difference is made to a thing θ by the presence or absence of a certain condition c, you know what θ is like with c, and also what θ is like without c, and on comparing the two find no difference. This involves knowing what θ is like without c; in the present case, knowing what you defined as the unknown. (A: 44–45)

In relation to presuppositionless knowledge, the argument can be reconstructed as stating that to claim that the world is independent from one's knowledge, one should have access to the world as one knows it, but also to the world independent from one's knowledge (D'Oro 2002: 42–43). By this, Collingwood rejects realism on the ground that it requires knowledge of things in themselves. The rejection of realism brings forth a comparison to pragmatism. D'Oro, for instance, emphasizes the link to pragmatism with respect to rejecting the correspondence theory of truth. However, despite his criticism of the correspondence theory of truth, unlike pragmatists, Collingwood does not seek to dissolve traditional metaphysical questions, but instead tries to provide a new way of investigating them. Also, unlike pragmatists, Collingwood does not abandon the normative character of philosophy (D'Oro 2002: 47). While D'Oro's discussion mainly refers to Rorty's version of pragmatism (see Rorty 1982, 1991), in Sect. 4 I will assess Collingwood's stance on the debate between realism and anti-realism, as well as his relation to different strands of pragmatism.

Regarding causation, Collingwood uses his method of investigating absolute presuppositions to identify the presupposition underlying causal thought (see EM ch. 3 and Collingwood 1937–38). According to Collingwood, this presupposition relies on the ability to distinguish between a present state of affairs which an agent aims to change into an intended state of affairs, "the *causa quod* [...] is a situation or state of things believed by the agent in question to exist' and 'the *causa ut*' is a purpose or state of things to be brought about" (EM: 292). This is also reflected in Collingwood's causation in sense I: "that which is 'caused' is the free and deliberate act of a conscious and responsible agent, and 'causing' him to do it means affording him a motive for doing it" (EM: 285). Collingwood deems sense I logically prior to the other two, that is, acting to bring about a certain effect—sense II, or causal interactions between events in nature—sense III (EM: 292, 323). This is in line with his earlier claim that absolute presuppositions are logically prior to questions (EM: 21), that is, to investigate what caused an event, one presupposes a relation between two events akin to that between an agent and the pursuit of a goal to hold. Notable for the metaphilosophical focus of this chapter, investigating causation through the absolute presuppositions underlying causal thought keeps the direction from epistemology to metaphysics: the meaning causation is presupposed to hold connects to the methods of inquiry. I explore this further in Sect. 5.

2 Collingwood and Pragmatism

Having sketched Collingwood's considerations on metaphysics, I will now explore the connection to pragmatism. My focus here will rest on more recent versions, as opposed classical pragmatism. This is first to emphasize the relevance of Collingwood's work for current developments of pragmatism and philosophy of science. Second, since there are several features specific to pragmatist thought, distinguishable in the work of contemporary authors associated with pragmatism, particular characteristics can be broken down for mapping out Collingwood's view. Particularly, I take features within Putnam's and Price's philosophy to match Collingwood's views more closely given the following broad shared characteristics: the denial of the correspondence theory of truth, the denial of distinctions such as appearance-reality or thought-experience, the denial of 'pure being' as the subject of philosophy and the focus on practices of inquiry. I will elaborate on particular issues regarding realism and causation in Sects. 4 and 5. For the purposes of the current section, I will also point out the links to classic pragmatist thought to provide a cohesive picture of pragmatism and its history, given the difficulty of finding a unifying definition.

In a narrow sense, pragmatism can be defined through the pragmatist maxim, traced to the classic pragmatist philosophers, Peirce, James, and Dewey: "a distinctive rule or method for becoming reflectively clear about the contents of concepts and hypotheses: we clarify a hypothesis by identifying its practical consequences" (Hookway 2016: 2). This maxim is not universally accepted among later philosophers associated with pragmatism, thus, a definition capturing a broader sense, would associate pragmatism with particular stances on classic philosophical topics and debates. For instance, Putnam singles out the following characteristics of pragmatism: (1) antiscepticism; (2) fallibilism; (3) rejecting the fundamental dichotomy between facts and values; and (4) the primacy of practice (Putnam 1995: 152). In Hookway's reconstruction, several other dichotomies are involved besides facts and values, such as mind-body, thought-experience, and analytic-synthetic (Hookway 2016: 5). Though Collingwood can be said to share with the classical understanding of pragmatism, the focus on method, as well as the aim of conceptual clarification, he does not focus directly on consequences. Rather, a certain concept or hypothesis is examined in the context of the absolute presuppositions underlying the process of inquiry where the given concepts or hypotheses belong. Note that this does connect to the usage of concepts, and specific epistemic practices, but unlike a view focusing on consequences, Collingwood aims to disclose the absolute presuppositions at work within the given practice.

Earlier pragmatist approaches to truth enabled later views such as Rorty's deflationism, which, as noted by D'Oro (2002), stands in contrast with Collingwood's philosophy. Notably, Rorty adopted a deflationist view on the majority of the traditional problems of philosophy, such as the debate between realism and anti-realism. While Collingwood also criticized truth as correspondence, he did not support anything akin to the pragmatist theory of truth. Thus, given its focus on critiquing traditional philosophical debates and distinctions, Rorty's strand of pragmatism appears to be taking the critique of metaphysics farther than Collingwood. Furthermore, Collingwood proposed the reform of metaphysics through an epistemological project, while Rorty proclaimed 'the death of epistemology'. In the light of these differences, I argue that Collingwood's views are much closer to pragmatist approaches with different focus, such as Putnam's. First, Putnam also rejects the pragmatist theory of truth (and the label of 'pragmatism' altogether) (see Rorty et al. 2004). Second, Collingwood's argument against realism fits the project of denying the dichotomy between appearance and reality. As I will argue in the next section, there are important similarities between Collingwood's case against realism, and Putnam's 'internal realism'. A final common point between Collingwood and pragmatist thought for my purposes here is the denial of an a priori metaphysics. I should note a connection with Peirce's considerations on the pragmatist maxim, understood as "a logical rule that shows the emptiness of 'concepts' which have no practical consequences" (Hookway 2016: 6), and thus reveal connections to classical pragmatism. Metaphysics is not concerned with a priori truths about 'pure being' for Collingwood either, but rather with absolute presuppositions, which are investigated in their historical contexts. The claim that certain philosophical problems can be addressed only by investigating human practices (as opposed to a study of 'pure being') is currently adopted by philosophers taking a pragmatist stance, such as Price, Brandom, and Kitcher. In Sect. 5, I will further explore the connection to Price, given the discussion of the common topic of causation.

These broad similarities between Collingwood's work and contemporary contributions to pragmatist thought, particularly by Putnam and Price will provide the framework for a more detailed investigation of pragmatist approaches to the issues of realism and causation. As both issues have been important topics in twentieth-century philosophy of science, the analysis will stress the importance of Collingwood's work within this field.

3 Collingwood and Putnam on Realism and Anti-Realism

In this section, I emphasize the similarities between Collingwood's and Putnam's considerations on realism. I focus on the distinction between metaphysical realism and internal realism, as well as on the role of Kant's philosophy as an earlier statement of the views above, and further on a broader metaphilosophical statement concerning the role of metaphysical inquiry. I argue that there are important similarities regarding Collingwood's and Putnam's methods of investigating realism, while the broader aims and philosophical views they endorse are distinct.

Putnam describes metaphysical realism, or externalism, as a view holding that the world is constituted by mind-independent objects (Putnam 1981: 49). According to externalism, it is possible to provide a complete description of the world as it is, from a perspective that Putnam deems a 'God's Eye point of view'. Putnam contrasts this view with internalism, defined as holding that multiple descriptions of the world are possible from different perspectives "of actual persons reflecting various interests and purposes that their descriptions and theories subserve", and associates this view with pragmatism (ibid.: 50). According to Putnam, externalism, despite being among the oldest views in philosophy, is untenable; he goes on to defend internalism, whose origins he traces back to Kant.

There are important similarities between Putnam's and Collingwood's views on realism, but there are also points of contrast. Both philosophers rejected the possibility of knowing the world from a God's Eye viewpoint. The plurality of actual viewpoints sketched by Putnam in terms of interests and purposes can also be explained in terms of absolute presuppositions. According to Collingwood's view, absolute presuppositions underlie scientific theories without being part of the theories themselves, or subject to evaluation with respect to truth value. The further step in bringing these similarities together is to hold that absolute presuppositions express interests and purposes specific to a given historical context of inquiry, which are not questioned among those conducting the inquiry, but subject to philosophical analysis.

Given these similarities, two important divergences should be noted. First, Putnam's description of realism starts off from an ontological perspective, the existence of mind-independent objects, and only later explores its epistemological consequences, that is, a version of the correspondence theory of truth. As explained above, Collingwood's argument against realism renders attempts to describe the world in terms of mind-independent objects impossible from the onset as such endeavours would assume knowledge of things in themselves. Unlike Putnam's ontological claims, Collingwood's argument against realism relies on an epistemological thesis—that it is impossible to know the world in itself. Second, while there are epistemological arguments on both sides, it should be noted that sceptical scenarios play an important role in Putnam's defence of internal realism. Namely, Putnam explains that the Brain in a Vat scenario can be readily dismissed by the internalist: in a world where there are brains in vats, one can observe whether other people are brains in vats, but envisioning a world where everyone is a brain in a vat would require a God's Eye view (ibid.: 50). This line of defence is not available to the externalist, who would require a justification independent of the individual observer's perspective. As the discussion above shows, this is a refutation of a global sceptical scenario. While scepticism was among the widely discussed topics during Collingwood's time, he did not focus on refuting scepticism (D'Oro 2002: 30). On Collingwood's view, realism does not hold because it rests on an assumption of presuppositionless knowledge, and the ability of answering sceptical arguments is not a decisive point for favouring one view over another.

A different kind of similarity stems from the influence of Kant's philosophy on both thinkers. Putnam identifies Kant as the first representative of internalism and further explores several Kantian themes in his works (Putnam 1981: 74). While Putnam openly acknowledges Kant's view as crucially important to his approach, the connection between Collingwood and Kantian philosophy is subject to debate (D'Oro and Connelly 2015: 5.3). For my purposes, here I rely on certain important Kantian themes in Collingwood's work identified by D'Oro: both philosophers' proposals for reforming metaphysics, their acknowledgement of a distinction between ontological and epistemological dimensions of idealism, and philosophical method—namely, explaining our experience in *An Essay on Metaphysics*, or its conditions of possibility in Kant's *Critique of Pure Reason* (D'Oro 2002: 25–28). These similarities show important convergences between the views of the two philosophers.

Comparing how Kant's philosophy influenced Collingwood and Putnam, the most important themes to stress are realism and the reform of metaphysics. I take the following claim by Putnam to be relevant: "if we do shift our way of thinking to the extent of regarding 'the world' as partly constituted by the representing mind, then many things in our popular philosophy [...] must be reexamined. [...] Kant held that [...] this *philosophical question* was never going to be solved by empirical science. I am suggesting that on this subject Kant was right" (Putnam 1992: 162). This passage showing Putnam's opposition to realism is remarkably similar to Collingwood's argument: the world being partly constituted by the mind entails that there is no mind-independent, that is, presuppositionless knowledge. Furthermore, convergence on the wider metaphilosophical stances of the two philosophers and Kant is revealed: the debate between realism and anti-realism cannot be solved by the means of science and belongs to the realm of metaphysics. This falls in line with Collingwood's claim that absolute presuppositions fall within the area of metaphysical investigation, and are different from empirical claims, for instance, by not having a truth value.

The considerations above provide an insight into the pragmatist topics within Collingwood's and Putnam's work. Directing the investigation towards philosophy of science, there is a question concerning realism as a matter of degree. While both Collingwood and Putnam deny a particular strand of realism that has dominated philosophy for centuries, so far I did not clarify the extent to which they reject realist assumptions. To answer this question, I rely on the view that commitments to realism come in degree, as contemporary discussions on objectivity and scientific realism have shown.[1] Putnam criticized externalism and the correspondence theory of truth, while at the same time supporting a different branch of realism, concerning theoretical entities (such as electrons). One of his well-known statements is that: "realism is the only philosophy that doesn't make the success of science a miracle" (Putnam 1975: 73). While Putnam is a critic of realism, he does not entirely dispense with realist tenets, particularly regarding theoretical entities in the sciences. Thus, if we view the philosophical stances between realism and anti-realism as a spectrum, Putnam's overall view would be closer to realism than Collingwood's. As Collingwood was not particularly concerned with theoretical entities, it would be difficult to reconstruct his stance on this issue for comparison with Putnam's view. A preliminary conclusion of the discussion above is that Collingwood and Putnam use comparable strategies in attacking realism, and trace their stance regarding metaphysics to similar historical sources. Nevertheless, their ultimate conclusions are different.

To sum up, Collingwood's work has similarities to Putnam's stance on pragmatism, namely, involvement in the debates over realism and anti-realism and particularly the critique of realism. While their contributions reflect rejections of particular dichotomies, notably between primary and secondary qualities, and appearance and reality, neither philosopher pursued dissolving philosophical problems as other strands of pragmatism proposed. Broadly, this fits the project of reforming metaphysics, or, more narrowly, the metaphysical debates over realism, and both philosophers converge on the view that the investigation of these issues is not empirical. While there are differences in their engagement with ontology, the problem of scepticism, and to what degree the two philosophers are realist, there are methodological similarities and shared historical influences. I take these similarities to be important especially given the metaphilosophical dimension of Collingwood's work and of pragmatism: the strategy he employed is largely compatible with Putnam's pragmatist approach. However, in the next section, I show how Collingwood's treatment of causation brings his view closer to Price's pragmatism than to Putnam's.

4 Collingwood and Price on Investigating Causal Concepts Through Human Practices

In the context of Collingwood's philosophy, the analysis of causation is an illustration of his method of investigating absolute presuppositions. The connection between causation and realism, examined in the previous section, should also be noted: Price, whose views I compare to Collingwood's, favours a projectivist view over realism: causation is said to be a secondary quality, tied to specific human capacities (see Price 1991; Menzies and Price 1993; Price 2007). Price (2001) presents this view in a pragmatist key and also points to a connection with Putnam's denial of the distinction between primary and secondary qualities, which is part of Putnam's broader critique of externalism described above. The difference to stress, however, is that Collingwood's discussion of the three senses of causation targets three types of causal explanation, and thus there is no ontological commitment regarding causation. As I will explain below, while Price discusses a version of ontological subjectivity, which would run against causal realism, he focuses on a practice-driven sense. Thus, insofar as neither

Collingwood, nor Price, commit themselves to realism, the similarity between their overall views appears to be stronger than in Putnam's case. In this section, I explore specific similarities regarding causation.

Price draws several distinctions with respect to subjectivity about causation, one of which refers to ontology, but focuses on one aspect: the practice-subjectivity of causation. He acknowledges this as a pragmatist stance: "the practice-subjective doctrine that we don't understand the notion of causation—as philosophers, as it were—until we understand its origins in the lives and practice of agents such as ourselves" (Price 2001: 107). He traces this strategy of analysing causation to Ramsey's (1978) considerations on probability, which, in Price's interpretation, cannot be characterized without its role in decision-making, which is a human endeavour (Price 2001: 107). Price distinguishes between global pragmatism and pragmatism about causation on the basis of a divergence between the realist doctrine about theoretical entities and causal realism. Briefly, his view goes as follows: "being realist about causation is not like being realist about electrons, or any of the other postulates of the special sciences. Causation is different, and it is pragmatism, not realism, which offers an account of the difference" (ibid.: 120). How does pragmatism explain the difference? According to Price, the role of causation in the sciences is different from that of theoretical entities: "causation is the medium in which the investigation is conducted, the thread which leads to unseen objects. Treat the thread itself as unknown, and we are simply blind" (ibid.: 114). Since the details of Price's argument go beyond the purpose of my chapter, I will comment on his overarching view.[2]

There are several analogies to draw between Price's project and Collingwood's approach. First, the subjective component of causation under Price's view, namely the agent's situation, is consistent with Collingwood's formulation of the absolute presupposition underlying causal thought, that is, a reason prompting a particular action. The more general presupposition rests on an explanation scheme involving *causa quod* (efficient cause) and *causa ut* (final cause) which are present in Collingwood's explanation of sense I causation—performing an action in pursuit of a goal. As specified in Sect. 1, according to Collingwood, this sense is prior to the other two senses—the causal connection between an action and its effect (sense II), or between events in nature (sense III). Thus, on Collingwood's view, notably, causal explanation in history is rendered possible by the presupposition that agents perceive the distinction between a current and a desired state of affairs and act to bring about the latter.[3] While causal explanation in history is central to Collingwood's work, the model can be extended to a broader claim that absolute presuppositions render specific forms of inquiry possible (as illustrated by causal explanations reliant on senses II and III). These tenets by Collingwood can be represented as standing in logical connection with Price's claim that causation is only understood by investigating the agent's practices, particularly decision-making. To clarify how Price's claim can be a consequence of Collingwood's global view:

1. Absolute presuppositions enable means of inquiry specific to particular sciences (Collingwood's definition of absolute presuppositions).
2. Absolute presuppositions are the object of study of philosophy (EPM, EM).
3. The absolute presupposition enabling causal thought consists of a distinction between a current and a desired state of affairs and the ability to act to enact the latter—the relation between *causa quod* and *causa ut* (Collingwood on causation).
4. The ability to act such as to turn a current state of affairs into a desired one (the relation between *causa quod* and *causa ut*) is characteristic of human agents and tied to human practices.
5. Therefore, a philosophical analysis of causal concepts in the context of scientific inquiry can only be conducted by investigating human agents' practices (Price's pragmatism about causation).

If this is correct, Price's view may be logically connected to an overall view on the presuppositions underlying causal thought of the kind Collingwood supports. To put it another way, questions concerning causation are enabled by a stance that takes into consideration an agent's actions and perspective. As specified above, while Collingwood's analysis connecting motives to actions mainly applies to history, causal thought is modelled on human practices in the case of other special sciences too, as Price's considerations on decision-making show. Further in-depth similarities between Collingwood's view and Price's perspectivalism are explored elsewhere (see Popa 2016: §4). While this point shows Collingwood's work to be relevant to contemporary debates on causation and agency, the connection to pragmatism will be more thoroughly established in what follows.

The other important point to make in this section is that the distinction between first-order entities in science and causation as a medium for the investigation made by Price in a pragmatist context can also be drawn using Collingwood's method. The project of investigating absolute presuppositions rests on a distinction between scientific knowledge and the presuppositions enabling the inquiry. On this view, causal connections are not akin to the entities scientists posit, but, rather, belong to the category of absolute presuppositions rendering the whole process of inquiry possible. To investigate the causal powers of electrons is to assume that looking for properties which enable an entity to cause certain events is a valid way of conducting research. For comparison, I refer to Collingwood's example concerning the absolute presupposition that events are caused:

> [I]f you were talking to a pathologist about a certain disease and asked him 'What is the cause of event E which you say sometimes happens in this disease?' he will reply 'The cause of E is C'; and if he were in a communicative mood he might go on to say 'That was established by So-and-so, in a piece of research that is now regarded as classical.' You might go on to ask 'I suppose before So-and-so found out what the cause of E was, he was quite sure it had a cause?'. The answer would be 'Quite sure, of course.' If you now ask 'Why?' he will probably reply 'Because everything that happens has a cause.' If you are importunate enough to ask 'But how do you know that everything that happens has a cause?' he will probably blow up right in your face, because you have put your finger on one of his absolute presuppositions, and people are apt to be ticklish in their absolute presuppositions. But if he keeps his temper and gives you a civil and candid answer it will be to the following effect. 'That is a thing we take for granted in my job. We don't question it. We don't try to verify it. It isn't a thing anybody has discovered, like microbes or the circulation of the blood. It is a thing we just take for granted.' (EM: 31).

As this passage shows, the role of causation is not to be sought among the first-order entities used in research on pathology. Rather, causation is part of the absolute presuppositions enabling the inquiry. This point is in line with Price's pragmatism regarding stipulating a different status for causation in comparison to first-order entities in the sciences. While no claims about what microbes bring about what disease can be made without a concept of causation, causation itself cannot be investigated by the methods of pathology. Moreover, Collingwood also mentions that this presupposition no longer holds in every area of science, which is also in line with Price's considerations on causation in the special sciences, and with further contemporary debates on whether there is causation in fundamental physics (see for instance (Field 2003)). It should be noted that this investigation can be extended to Price's overall pragmatist view involving 'practical stances', causation being one instance among examples such as probabilities and conditionals (Price 2013: 47–48). Insofar as absolute presuppositions underlie various forms of inquiry, causal thought alongside other means of investigation enable various scientific fields without being part of the content of the said sciences.[4]

In the light of the point above, the affinity between Collingwood's view on causal explanation and Price's pragmatism can be further emphasized. Both authors distinguish between first-order claims of specific sciences and the frame of inquiry, and on both views, causal concepts can be fully analysed only by reference to the human practices constituting the frame. Thus, when causal concepts are used in explanations in the special sciences, they inevitably reflect aspects of the human subject. This point is emphasized by Price when discussing Putnam's interpretation of Kant and instrumentalism—causal explanations are significant from the perspective of human concerns rather than by reference to a mind-independent reality. Thus, Collingwood and Price converge on a view on causation as fundamentally connected to human practices and causal explanation is always set against a corresponding frame of inquiry. Price (2001: 108) holds that if there were no thinking beings there would be no causal concepts, and, by extension, no search for causal explanations.

In summary, there are notable affinities between Collingwood's method of investigating absolute presuppositions together with his specific view on causation and Price's pragmatism. First, understanding causation through an absolute presupposition that refers to a human capacity (acting in pursuit of goals) entails a view on causal understanding along the lines of Price's practice-subjectivity—that is, we can understand causation only by looking into its origins in particular human practices. Second, the role of causation in scientific inquiry is similar within Collingwood's method of investigating absolute presuppositions and Price's pragmatist framework: causation is not part of the set of entities scientists postulate, but part of the broader frame of inquiry, which enables research in a given field. The resulting causal explanations are dependent on the corresponding frame on both views.

5 Collingwood's Pragmatism

Having mapped out different strands of pragmatism and highlighted similarities between Collingwood's thought and contributions to pragmatism by Putnam and Price, I will end by placing Collingwood's view among the aspects of pragmatism discussed so far. This will enable an assessment of the extent to which Collingwood's philosophy can be classified as pragmatist. In the light of the discussion above, this section holds that Collingwood's view is closer to Price's with regard to the considerations on a frame of inquiry and causal explanation. At the same time, there are notable affinities between Collingwood's views and Putnam's contributions to pragmatism with regard to ways of contesting metaphysical realism and traditional philosophical dichotomies which are also present in Price's pragmatism.

Broadly, Collingwood's critique of realism and the correspondence theory of truth is consistent with the multiple strands of pragmatism explored here. Nevertheless, the proposed reform of metaphysics sets Collingwood apart from a pragmatist theory of truth and the denial of traditional philosophical debates as in Rorty's pragmatism, while his reliance on epistemology as opposed to ontology excludes views such as Putnam's internal realism. In terms of placing Collingwood's view on a spectrum between realism and anti-realism, in the light of the present analysis, Price's pragmatism about causation appears to be the closest. The analysis of causal concepts in the context of human practices parallels Collingwood's statement of the absolute presupposition underlying causal thought, and placing causation within the frame as opposed to the content of inquiry. A further common point emerges when considering the connection between the frame of inquiry and causal explanations: causal explanations and uses of causal concepts generally are dependent on the frame as opposed to referring exclusively to mind-independent entities. In Collingwood's case, this frame is constituted by absolute presuppositions, while Price distinguishes between the investigation itself and its medium with causation belonging to the latter. Nevertheless, the views converge as far as pragmatism about causal concepts and the rejection of realism is concerned. A question arising here is whether Collingwood's considerations on absolute presuppositions may be closer to a kind of global pragmatism as opposed to Price's pragmatism about causation. Noting that an answer to this question would require further investigation of certain claims belonging to scientific realism that Price refers to, such as entity realism, I restrict my comparison to a broad feature of pragmatism—the reference to human practices—and the resulting view on philosophical concepts such as causation. As causation was one particular illustration of Collingwood's method of investigating absolute presuppositions, pragmatist analyses of further concepts are possible, without necessarily committing to global pragmatism.

Having noted the pragmatist treatment of causation in the case of both Price and Collingwood, alongside the critique of realism, one final point to note is that much of Price's discussion relies on Putnam's contributions to pragmatism, particularly concerning the denial of philosophical dichotomies and contributions regarding instrumentalism. Thus, despite the divergences concerning realism and ontology, the methodological similarities between Collingwood's investigation of absolute presuppositions and Putnam's and Price's pragmatism should be emphasized. The denial of a 'God's eye' perspective on metaphysics and presuppositionless knowledge enabled similar critical stances towards dichotomies such as appearance and reality, or primary and secondary qualities, and thus a pragmatist outlook on philosophy.

Thus, the connections between Collingwood's views and the strands of pragmatism discussed here are twofold. First, in terms of distinguishing between a frame of inquiry and first-order scientific knowledge and articulating a pragmatist view on causal concepts with the potential of providing similar analyses to other concepts, Collingwood's view is closer to Price's. Second, a focus on methodology reveals similarities to Putnam's critique of metaphysical realism, and a pragmatist viewpoint on important philosophical debates, with the notable exceptions of scepticism and ontological inquiry. This shows a complex picture of affinities along different aspects of Collingwood's work and different strands of pragmatism.

6 Conclusion: Collingwood and Pragmatist Approaches to Realism and Causation

This chapter has explored similarities between Collingwood's overall method of investigating presuppositions applied to the issues of realism and causation, and current strands of pragmatism by Putnam and Price. Analysing Collingwood's critique of realism in comparison to Putnam's, I noted that both philosophers denied certain traditional philosophical dichotomies (thought vs. experience, primary vs. secondary qualities, appearance vs. reality, etc.), while being participants in debates concerning others (realism vs. anti-realism). From a methodological point of view, both philosophers took the problem of realism to be the subject of philosophy, as opposed to empirical science. This points to deep methodological similarities between the two thinkers and emphasizes the compatibility between Collingwood's approach and the pragmatist strategy. This point is further strengthened by the divergences between Putnam and Collingwood concerning arguments against realism and Putnam's focus on scepticism, and the ultimately different conclusions regarding realism—Putnam's critique is meant to support a different kind of realism, while Collingwood's critique is meant to establish that the claim that the world is independent of the subject's knowledge cannot hold, given the impossibility of presuppositionless knowledge. While the two philosophers pursued different aims, the methods they employed led them to views marked by some of the main tenets of pragmatism.

The investigation of Price's pragmatist stance on causation and Collingwood's considerations on the absolute presupposition underlying causal thought has shown similarities regarding the methodology, as well as concerning the view on the role of causation in scientific inquiry. On both authors' accounts, causation is to be investigated in the context of specific human practices from which causal thought arises, particularly in connection to agency. Likewise, on both views, there is a distinction between the main tenets of science and the tenets and concepts that shape the process of inquiry, and causation is part of the latter. I take this final point to be the main point of convergence between Collingwood and the pragmatist views explored here: current debates in the philosophy of science have involved distinctions between first-order entities and a certain frame, or set of conditions for the inquiry. As pointed out above, a similar strategy was employed by Collingwood in his suggestion of reforming metaphysics. The subsequent debates within the philosophy of science, especially those pertaining to pragmatism, emphasize the importance of drawing the distinction in the first place, and the need for further research on methodology: how do specific methods and practices affect our understanding of concepts used in the sciences? In this respect, Collingwood's work on causation deserves further emphasis, both through its focus on action, which is gaining wide acceptability in current debates, as well as through the epistemically driven method: one can find out what causation is by studying one's means of investigating causal connections.

From the broader perspective of contemporary debates on metaphilosophy, the current chapter has shown that Collingwood's method is comparable to particular pragmatist approaches: Putnam's critique of metaphysics and Price's considerations on causation, a frame of inquiry, and human concerns. While Collingwood does not subscribe to a fully deflationist view, his proposed reform of metaphysics does away with certain traditional distinctions and assumptions. In the light of the argument here, Collingwood's view can be said to have anticipated metaphysical debates concerning scientific concepts and practices of inquiry and to have provided a method that can be used in addressing some of these issues.[5]

Notes

1. See (Douglas 2004) for objectivity and (Chakravartty 2017) for scientific realism.
2. Briefly put, Price argues that while the existence of theoretical entities is proved by their causal interactions, the same move would not hold for causation, since that would be circular.
3. This interpretation would bring together *An Essay on Metaphysics* and *The Idea of History*.
4. An analogy can be drawn here to D'Oro's comparison between Collingwood's critique of metaphysics and Carnap's considerations on the internal/external distinction. Carnap criticizes metaphysics for aiming at answering external questions while questions in general can only be answered within a frame. Collingwood's view supports this distinction, but shifts the object of metaphysics to the frame itself (D'Oro 2015).
5. I am grateful to Giuseppina D'Oro and Karim Dharamsi for comments that helped strengthen this chapter and to Stephen Leach for editorial suggestions.

Bibliography

Boucher, David, and Teresa Smith, eds. 2013. *R.G. Collingwood: An Autobiography and Other Writings: With Essays on Collingwood's Life and Work*. Oxford: Oxford University Press.

Chakravartty, Anjan. 2017 Summer Edition. Scientific Realism. *The Stanford Encyclopedia of Philosophy*, ed. Edward N. Zalta. https://plato.stanford.edu/archives/sum2017/entries/scientific-realism/

Collingwood, R.G. 1937–38. On the So-Called Idea of Causation. *Proceedings of the Aristotelian Society* 38: 85–112.

———. 1993 [1946]. *The Idea of History*, Rev. ed. Oxford: Oxford University Press.

———. 1998 [1940]. *An Essay on Metaphysics*, Rev. ed. Oxford: Oxford University Press.

———. 2005 [1933]. An Essay on Philosophical Method, Rev. ed. Oxford: Oxford University Press.

D'Oro, Giuseppina. 2002. *Collingwood and the Metaphysics of Experience*. Oxford: Routledge.

———. 2015. Unlikely Bedfellows? Collingwood, Carnap and the Internal/External Distinction. *British Journal for the History of Philosophy* 23 (4): 802–817.

D'Oro, Giuseppina and James Connelly. 2015 Summer Edition. Robin George Collingwood. *The Stanford Encyclopedia of Philosophy*, ed. Edward N. Zalta. http://plato.stanford.edu/archives/sum2015/entries/collingwood/

Douglas, Heather. 2004. The Irreducible Complexity of Objectivity. *Synthese* 138 (3): 453–473.

Field, Hartry. 2003. Causation in a Physical World. In *Oxford Handbook of Metaphysics*, ed. Michael J. Loux and Dean W. Zimmerman, 435–460. Oxford: Oxford University Press.

Hookway, Christopher. 2016 Summer Edition. Pragmatism. *The Stanford Encyclopedia of Philosophy*, ed. Edward N. Zalta. https://plato.stanford.edu/archives/sum2016/entries/pragmatism/

Kant, Immanuel. 1998. *Critique of Pure Reason*. Cambridge: Cambridge University Press.

Martin, Rex. 1998. Editor's Introduction, to R. G. Collingwood's. *An Essay on Metaphysics*. Oxford: Oxford University Press.

Menzies, Peter, and Huw Price. 1993. Causation as a Secondary Quality. *The British Journal for the Philosophy of Science* 44 (2): 187–203.

Popa, Elena. 2016. Collingwood and Manipulability-Based Approaches to Causation: Methodological Issues. *Collingwood and British Idealism Studies* 22 (1): 139–166.

Price, Huw. 1991. Agency and Probabilistic Causality. *The British Journal for the Philosophy of Science* 42 (2): 157–176.

———. 2001. Causation in the Special Sciences the Case for Pragmatism. In *Stochastic Causality*, ed. Domenico Costantini, Maria Carla Galavotti, and Patrick Suppes, 103–120. Chicago: University of Chicago Press (CSLI Publications).

———. 2007. Causal Perspectivalism. In *Causation, Physics, and the Constitution of Reality: Russell's Republic Revisited*, ed. Huw Price and Richard Corry, 250–292. Oxford: Oxford University Press.

———. 2013. *Expressivism, Pragmatism, and Representationalism*. Cambridge: Cambridge University Press.

Putnam, Hilary. 1975. *Philosophical Papers: Vol. 1, Mathematics, Matter and Method*. Vol. 1. Cambridge: Cambridge University Press.

———. 1981. *Reason, Truth and History*. Vol. 3. Cambridge: Cambridge University Press.

———. 1992. *Realism with a Human Face*. Cambridge, MA: Harvard University Press.

———. 1995. *Words and Life*. New ed. Cambridge, MA: Harvard University Press.

Ramsey, Frank Plumpton. 1978. General Propositions and Causality. In *Foundations: Essays in Philosophy, Logic, Mathematics and Economics*, ed. D.H. Mellor, 133–151. London: Routledge & Kegan Paul.

Rorty, Richard. 1982. *Consequences of Pragmatism: Essays, 1972–1980*. Minneapolis: University of Minnesota Press.

———. 1991. *Objectivity Relativism, and Truth: Philosophical Papers*. Vol. 1. Cambridge: Cambridge University Press.

Rorty, Richard, Hilary Putnam, James Conant, and Gretchen Helfrich. 2004. What Is Pragmatism? *Think* 3 (8): 71–88.

CHAPTER 7

Why Epistemic Pluralism Does not Entail Relativism: Collingwood's Hinge Epistemology

Giuseppina D'Oro

1 Introduction

There is a widespread view according to which the denial that the conditions of knowledge are truth-evaluable inevitably leads to a form of epistemic pluralism that is *both* quietist *and* internally incoherent. It is quietist because it undermines the possibility of genuine epistemic disagreement. It is internally incoherent because it simultaneously denies the existence of universal knowledge claims and makes the universal claim that there is no such knowledge. The goal of this chapter is to show that denying that the conditions of knowledge are truth-evaluable does not necessarily entail a commitment to a form of epistemic relativism that is both quietist and internally incoherent. To undermine the view that the denial that the conditions of knowledge have truth-values leads down the blind alley of epistemic relativism I mobilize a version of "hinge epistemology" which distinguishes between epistemic pluralism and epistemic relativism. This form of hinge epistemology is to be found in Collingwood's account of

G. D'Oro (✉)
School of Politics, IR and Philosophy, Keele University, Staffordshire, UK
e-mail: g.d'oro@keele.ac.uk

K. Dharamsi et al. (eds.), *Collingwood on Philosophical Methodology*, Philosophers in Depth,
https://doi.org/10.1007/978-3-030-02432-1_7

absolute presuppositions (EM). By teasing apart epistemic pluralism from epistemic relativism, the chapter exposes the view that the denial that the conditions of knowledge have truth-values inevitably leads to a malignant form of epistemic relativism as a form of philosophical scaremongering.

The chapter begins, in Sect. 1, by considering Boghossian's characterization of epistemic pluralism in *Fear of Knowledge* (Boghossian 2006). According to Boghossian, the descriptive claim that there are different belief systems, combined with the denial that the conditions of knowledge are truth-evaluable, leads to a questionable form of epistemic pluralism, one which is relativist in outlook. I consider Boghossian's account of epistemic pluralism because it seems to capture the widespread view that pluralism and relativism must go hand in hand if one denies that the conditions of knowledge have truth-values. Section 2 outlines a form of epistemic pluralism that, I argue, is unfairly described as relativistic. This form of non-relativistic pluralism arises not in response to the descriptive claim that there is a plurality of belief systems, but to the normative claim that explanation should be fit for purpose. Once pluralism is conceived in this light it no longer has the quietest overtones that tend to be characteristic of epistemic relativism. The distinction between the relativistic pluralism that is the target of Boghossian's critique and non-relativistic pluralism is illustrated through a reconstruction of Collingwood's account of absolute presuppositions in *An Essay on Metaphysics*, a highly neglected but important contribution to hinge epistemology.[1] This reconstruction is offered as an illustration of how hinge epistemology could be construed to avoid two standard objections that are often raised against epistemic relativism, namely, that it makes it impossible to criticize other cultures and that it is self-undermining. Section 3 considers these two standard objections and argues that they do not apply to the kind of epistemic pluralism that arises from the consideration that explanation must be fit for purpose and sensitive to the goals of inquiry. The decoupling of epistemic pluralism from epistemic relativism argued for in this chapter rests on a reading of the role of absolute presuppositions as non-truth-evaluable epistemic hinges that is only presented rather than defended here. I have defended this reading elsewhere (D'Oro 2002, 2010), but the argument of this chapter does not depend on whether the ascription of this hinge epistemology to Collingwood is factually correct, but on whether, given the way it is presented here, it succeeds in decoupling epistemic pluralism from epistemic relativism and in defending the former from the charges that are levelled against the latter.

2 Epistemic Pluralism and the Charge of Relativism

Boghossian defines epistemic pluralism as follows:

> (i) "there are many fundamentally different, genuinely alternative epistemic systems, but
> (ii) no facts by virtue of which one of these systems is more correct than any of the others' (Boghossian 2006: 73).

(i) is an uncontroversial descriptive claim about the existence of different belief systems; (ii) is a controversial claim that pluralists must be committed to if, unlike absolutists, they deny that belief systems have truth-values. When (ii) is taken in conjunction with (i) it turns the benign descriptive pluralism of (i) into a malignant form of relativistic pluralism. For if different belief systems cannot be assessed for truth and falsity, then what is true/false is relative to the method of justification which belongs to a particular belief system and it is therefore impossible "to speak of what is rational period, but only of what is rational relative to this or that accepted epistemic system" (Boghossian 2006: 63). Since hinge epistemologists deny that the notions of truth and falsity apply to the fundamental hinges on which knowledge rests, they are committed not only to (i) but also (ii) (Carter 2017). Boghossian illustrates the kind of epistemic relativism that follows from the endorsement of (i) in conjunction with (ii) by considering Rorty's (1981) discussion of the dispute between Cardinal Bellarmine and Galileo over the truth of heliocentrism. Both justified their conclusions by appeal to evidence, Galileo to the evidence of the telescope, and Cardinal Bellarmine to the evidence of the Bible. The Copernican, heliocentric universe is justified by Galileo's observations relative to the belief system of science. The Ptolemaic geocentric universe is justified by Bellarmine's reading of the Bible, relative to the belief system of the Catholic faith. Rorty concludes that since there is no fact of the matter as to whether the universe is Ptolemaic or Copernican, then describing Cardinal Bellarmine's view as unscientific is just a form of "sophisticated name calling" (Boghossian 2006: 62). Rorty's discussion of the dispute between

Bellarmine and Galileo, so Boghossian argues, echoes Wittgenstein's handling of the conflict between the Azande's pre-scientific culture, with its reliance on the oracle, and a scientific culture with its reliance on the propositions of physics. Just as Rorty regards any attempt critically to evaluate the epistemic claims of Bellarmine's views as a form of sophisticated name calling, so (Boghossian continues) Wittgenstein claims that: "Where two principles really do meet which cannot be reconciled with one another, then each man declares the other a fool and a heretic" (Wittgenstein 1969: §611). The moral of the story seems to be that if you are an epistemic pluralist then your options are either to remain silent or to exchange insults. There is no such thing as respectful disagreement. For the purposes of this chapter, I shall take Boghossian's reconstruction of Rorty's relativism at face value and turn to the following question: is epistemic pluralism necessarily relativistic in outlook? In the next section, I outline a kind of hinge epistemology which is epistemically pluralist and yet not relativist. I argue that this form of hinge epistemology denies that the conditions of knowledge are truth-evaluable but is not committed to the objectionable relativism which, according to Boghossian, entails that one cannot disagree with the epistemic claims of other cultures. This form of epistemic pluralism without the controversial relativism that is normally associated with it is articulated in Collingwood's account of absolute presuppositions in *An Essay on Metaphysics*, which I present as making an important, if neglected, contribution to hinge epistemology (the view that the fundamental hinges on which knowledge turns are not truth-evaluable).

3 Collingwood's Hinge Epistemology

The project of hinge epistemology has undergone a revival in the wake of the attention that has lately been given to Wittgenstein's *On Certainty*. (Moyal-Sharrock 2004; Moyal-Sharrock and Brenner 2007; Hamilton 2014). While the affinities between Collingwood's account of absolute presuppositions in *An Essay on Metaphysics* and Wittgenstein's account of hinge propositions in *On Certainty* have not gone unnoticed (Festin 2009), the former has not received much attention in the recent and growing body of literature on hinge epistemology (Coliva 2010, 2015). This section introduces Collingwood's version of hinge epistemology with a view to disentangling epistemic pluralism from epistemic relativism and showing that the denial that the conditions of knowledge have truth-values does not necessarily undermine the possibility of epistemic disagreement.

3.1 Absolute Presuppositions as Hinges

In *An Essay on Metaphysics* Collingwood argued that there are certain propositions to which the notion of truth and falsity does not apply. In fact, these so-called propositions are not genuine propositions at all precisely because, unlike ordinary propositions, the notion of truth and falsity does not apply to them. The principle of the uniformity of nature, for example, has the grammatical form of a proposition, but it should not be interpreted as one. (EM: 152) Collingwood gave these apparent propositions the technical name of "absolute presuppositions". He denied that absolute presuppositions are propositions in the ordinary sense and claimed that metaphysics, properly understood, is concerned not with propositions, but with absolute presuppositions. Since the notion of truth and falsity does not apply to absolute presuppositions, these are *beyond* verification, not in the sense in which, according to Ayer or Hume, metaphysical propositions such as "God exists" cannot be verified (either empirically or analytically), but because the notion of verifiability does not apply to them. Absolute presuppositions are therefore best described *not as being unverifiable* (lest this suggest that if they were to be put through the verificationist test they would fail it), *but as not being the sort of thing to which the notion of verification should be applied.* To bring to the fore the full force of the claim that the notion of verifiability does not apply to absolute presuppositions, one must say more about the distinction between propositions and presuppositions and how absolute presuppositions, which Collingwood takes to be the proper object of philosophical enquiry, differ from what he calls "relative presuppositions".

3.2 Propositions and Presuppositions

Propositions and presuppositions play very distinctive roles in what Collingwood called the logic of question and answer. A proposition's role is to answer a question. A presupposition's role is to give rise to questions. For example, "the salt is on the kitchen table" is a propositional answer to the question "where is the salt?" and this question relies on presupposition such as that there is salt, that there is a kitchen and so on. Collingwood's claim is that the ability of presuppositions to give rise to questions (their logical efficacy) has nothing to do with their being true or even being believed to be true/being accepted as true. The presupposition "Santa exists", for example, must be operative for a question such as "what did

Santa leave in your Christmas stocking?" to be asked, even if the proposition "Santa exists" is false. Presuppositions, therefore, do not do their logical work *in virtue of being true* since they have logical efficacy even if they are false (as in the case of Santa's existence). Nor do presuppositions do their logical work *in virtue of being believed to be true or being accepted as true*: the question "What did Santa bring you for Christmas?" entails the presupposition "Santa exists" whether it is asked by a child interrogating another child, or by an adult interrogating a child, even if the former believes in Santa and the latter does not. It is therefore not in virtue of *being accepted as true* or *believed* to be true that the presupposition "Santa exists" gives rise to the question "What did he bring for Christmas?" Presuppositions are therefore neither true propositions nor propositions which are believed to be true or accepted as true. As Collingwood says the distinction between presuppositions and propositions is "a matter of common knowledge" since "it is possible and often profitable to argue from suppositions which we know to be false, or concerning which we have neither knowledge nor belief as to whether they are true or false" (EM: 28).

It is sometimes possible for one and the same statement (to use an admittedly inadequate neutral expression) to occupy more than one role, that is, to operate either as a proposition or as a presupposition. In this case, the notion of verifiability will apply to it in one context but not in another. For example, when surveying a property with a tape measure one will presuppose that the distance between two walls will be the one recorded by the measuring tape. In this case, the accuracy of the tape will be presupposed. If one were to suspect that the measuring tape had stretched through time or wear and tear, one would treat the statement "the measuring tape is 66 foot long" (EM: 29) as one that could be found to be correct or incorrect and thus relate to it not as a presupposition, but as a proposition. What determines whether a statement is a proposition or a presupposition is not what it states, but the role that it plays in what Collingwood calls the logic of question and answer. Propositions are statements whose role is to answer questions. Presuppositions are statements which give rise to questions. Attempting to verify a presupposition by asking whether it is true or false is therefore a self-stultifying exercise because in treating a presupposition as something that can be true or false one automatically relates to it as a proposition. It would be analogous to using a knight as a bishop in the game of chess: just as what defines the knight, or the bishop, is how they move on the chessboard, so likewise what determines whether something is a proposition or a presupposition, is the role it plays in the asking of questions and giving of answers. The fact that some statements can play different roles, however, says something significant about them, namely, that when they play the role of presuppositions, they are only relatively presupposed. As Collingwood says:

> To question a presupposition is to demand that it should be 'verified'; that is, to demand that a question should be asked to which the affirmative answer would be that presupposition itself, now in the form of a proposition. To verify the presupposition that my measuring-tape is accurate is to ask a question admitting of the alternative answers 'the tape is accurate', 'the tape is not accurate'. Hence to speak of verifying a presupposition involves supposing that it is a relative presupposition. (EM: 30)

Statements which can be either presupposed or asserted as propositions Collingwood calls relative presuppositions and contrasts them with presuppositions which are absolute rather than relative.[2]

3.3 The Absoluteness of Absolute Presuppositions

Absolute presuppositions, Collingwood claims, have only one role in the logic of question and answer. "Everything that happens has a cause" is a presupposition that Collingwood regards as absolute rather than relative. He illustrates the absolute nature of this presupposition by considering how a pathologist might respond to an enquirer who asks questions concerning the cause of a disease and is given answers which appeal to a certain piece of research which has established what the cause of the said disease is. If the enquirer were not content with the answers and continued by asking "But how do you know that everything that happens has a cause?" the answer he would be likely to receive would be: "That is a thing we take for granted in my job. We don't question it. We don't try to verify it. It isn't a thing anybody has discovered, like microbes or the circulation of the blood. It is a thing we just take for granted" (EM: 31).

The pathologist's reply here ("That is a thing we take for granted in my job. We don't question it") might convey the misleading impression that absolute presuppositions are beliefs which are very entrenched and held dogmatically and thus exempted from critical scrutiny. While Collingwood denies that absolute presuppositions are answers to questions, it is not their going unquestioned that accounts for their special logical status. The point is not so much that it does not occur to the pathologist to question whether everything which happens has a cause, but rather that since the specific questions asked by the pathologist (What is the cause of malaria? What is the cause of cancer?) presuppose "everything which happens has a cause", such questions cannot be answered without invoking the presupposition which gives rise to those very questions. Since true as well as false answers to the question "What is the cause of cancer?" presuppose that everything, including cancer, has a cause, it makes no sense *both* to ask "what is the cause of this? What is the cause of that?" and to question whether everything which happens has a cause. As he says:

> any question involving the presupposition that an absolute presupposition is a proposition, such as the question 'Is it true?' 'What evidence is there for it? 'How can it be demonstrated?' 'What right have we to presuppose it if it can't?' is a nonsense question. (EM: 33)

Absolute presuppositions are therefore not propositions which are accepted as true dogmatically or unreflectively. They are unlike "The Earth is flat", which was widely believed to be true and went unquestioned for long periods of time. A presupposition is absolute, not if it goes unquestioned, but if it cannot be put in question without undermining the form of inquiry it undergirds. "The Earth is flat" is more like "water is H2O", which nobody (now) questions and everybody believes to be true, but which could conceivably be discovered to be false without jeopardizing the method by which the claim is verified. Absolute presuppositions are therefore not beliefs which are dogmatically adhered to without question. The unwillingness critically to question a presupposition is not a test of its absolute nature. The critical test for deciding whether a presupposition is absolute is not whether it goes unchallenged, but whether it makes sense to ask for verification in the first place.

Another possible interpretation of the absoluteness of absolute presuppositions is that they are like Kantian postulates in the sense that they are propositions which are necessarily believed to be true, even if they may not be. As presuppositions generally do not do their logical work in virtue of being *believed to be true*, likewise *absolute* presuppositions do not do their logical work in virtue of being *necessarily believed to be true*. Absolute presuppositions are therefore unlike Kantian postulates such as the freedom of the will or the immortality of the soul about which one can have no knowledge, but which one must hold as true in order to act morally. It is rather that it makes no sense to speak of knowing them to be either true or false because they are not propositions with true or false values. The term "absolute" may also misleadingly suggests that what makes presuppositions absolute is that they are necessary and universal in the manner of analytic truths. This is not the sense in which presuppositions are absolute. What makes the presupposition "everything which happens has a cause" absolute is that it is constitutive of the kind of investigative practice which the pathologist pursues. Without presupposing it the pathologist could not verify specific empirical correlations, such as those holding between smoking and lung cancer, or excessive drinking and liver disease. But a presupposition that is constitutive of one form of inquiry may not be constitutive of another. The principle of the uniformity of nature, for example, is a constitutive principle for the empirical investigation of reality. Without such a presupposition, it would not be possible to predict that ice in the Arctic will melt if the temperature rises because the empirical law that ice melts above 0°C relies on the assumption that past regularities provide a guide to future ones. Nor would it be possible to retrodict that dinosaurs disappeared in the Earth's geological past because they froze to death when the average temperature dropped very drastically. Such a presupposition, on the other hand, would be of no help in a different investigative context. For example, while the principle of the uniformity of nature plays an essential role in the scientific investigation of nature, it is not a principle that governs the historical concern with different cultures. It is an appeal to variations rather than uniformity in epistemic norms that enables humanistically oriented historians to explain that Tutankhamun was born of out of an incestuous union because the Ancient Egyptians believed this was a way of preserving the purity of the blood. What makes a presupposition absolute is not its strict universality, but its indispensability to a context of inquiry; the critical test for deciding whether a presupposition is absolute is *not* whether it is a strictly universal proposition whose denial generates a logical contradiction, but whether denying it entails a form of intellectual hypocrisy on behalf of those who invoke it in their explanatory practices. Absolute presuppositions are therefore not universal in the way in which necessary analytic truths such as "bachelors are unamarried" are universal. Analytic truths cannot be denied without contradiction, but there is no contradiction in asserting that the future will not resemble the past.

In sum, I have discarded three possible ways in which the absoluteness of absolute presuppositions might be construed. The first is the view that absolute presuppositions are beliefs which people unreflectively accept as true and exempt from critical scrutiny and thus that their "absoluteness" lies in their being dogmatically held. The presupposition "everything which happens has a cause" is *absolute* not because it goes questioned, but because it is necessarily presupposed by the pathologist's explanatory practice. It is not the pathologist's unquestioningly or dogmatically *believing* this presupposition to be true that makes the presupposition absolute: presuppositions do not do their logical work in virtue of being believed to be true. Second, I have rejected the view that a presupposition is absolute because it is necessary to believe it to be true (like a Kantian postulate) even if *ex hypothesi* it could turn out to be false. Finally, the absolute nature of presuppositions should not be conflated with the notion of strict universality because absolute presuppositions do their logical work not in so far as they are self-evident analytic truths, rather than, say, general empirical propositions, but in so far as they are invoked in the practices they undergird.

3.4 Philosophy's Role

In denying that the notion of truth and falsity applies to absolute presuppositions Collingwood signals his refusal to take part in the metaphysical dispute between ontological realists, who assert, for example, that propositions such as "nature is uniform" or "everything which happens has a cause" are true, and sceptical idealists who claim they are false. Traditional metaphysicians, who state that causes are real in the context of a mind-independent reality or that they are ideal, treat absolute presuppositions as if they were propositions with a definite truth value. But the task of metaphysics in Collingwood's view is not to expand knowledge by adding to the body of true propositions; it is to work backwards from the propositional answers which are produced in the sciences to the questions which they answer, and from these to the presuppositions which give rise to those questions, until one reaches presuppositions which are essential to that form of knowing: "The metaphysician's business … is not to propound them [absolute presuppositions] but to propound the proposition that this or that one of them is presupposed" (EM: 33). Collingwood's goal is not to settle metaphysical disputes, but to direct philosophical analysis away from such disputes and find a new focus for philosophical analysis: the study of absolute presuppositions. The task of philosophy is to uncover the presuppositions which give rise to the questions characteristic of different forms of inquiry, questions which receive propositional true or false answers in the special sciences. The uncovering of absolute presuppositions involves a logical regress from answers to the questions to which they are answers and from these questions to the presuppositions which gave rise to them. Such a task is not a matter of describing what system of beliefs people accept; for just as logic is concerned not with how people think, but with how they ought to think, likewise the regress from answers to questions and from questions to presuppositions uncovers not what people believe or what they accept as true, but what they presuppose.

When tracing the entailment relations which lead from answers to questions, and to presuppositions, one may find a mismatch between the answers given and the questions asked. Collingwood illustrates this point by using the example of a person whose car has stopped as he was driving up a steep hill and who wonders what the cause of the stoppage is (EM: 302–03). This person's question would not be answered by a theoretical physicist who pointed out that since the top of the hill is further removed from the centre of the Earth than the bottom of the hill the car requires more power to go uphill than downhill. It would be answered on the other hand by the AA person who discovered a loose cable and suggested that consideration as an explanation. The explanations of the theoretical physicist and of the car mechanic answer different kinds of why questions, questions which arise from different absolute presuppositions. The explanation of the theoretical physicist answers a question which presupposes an unconditional notion of causality; the explanation of the car mechanic answers a question which presupposes a conception of causes as handles. The point here is that these explanations are answers to different questions, which are entailed by different absolute presuppositions. The choice between these absolute presuppositions depends not on which one of them is true of an inquiry-independent fact, but on which one of them gives rise to the question that is satisfactorily answered by the car mechanic rather than the theoretical physicist.[3] Such is the nature of Collingwood's epistemic pluralism.

To illustrate this form of non-relativistic pluralism with a contemporary example, we might consider the case of the Russian ex-spy Sergei Skripal who collapsed in Salisbury outside a supermarket. What happened? If the "fact" to be explained is a physiological one, the answer will be that it was caused by exposure to the nerve agent Novichok. But if the fact to be explained is of a political nature, then knowing that it was nerve agent smeared on the handle of his front door that was responsible for his collapse, and that the nerve agent was Novichok, will at best enable the investigative journalist to restrict the list of suspects to those who have access to this rare chemical compound; it will contribute little or nothing to understanding the political significance of this fact. Explanations which appeal to antecedent conditions and those which appeal to motives do not conflict because they explain different facts, facts which are verifiable in very different ways. If the relevant fact is a physiological phenomenon, then it will have a physiological explanation; but if the relevant fact is of a political nature, then it will not satisfy the curiosity of those who want to know what happened in Salisbury to be told in reply that Sergei Skripal had made physical contact with the nerve agent Novichok, which had been smeared on the handle of his front door. The evidence that the physiologist and the political historian will produce to support their respective explanations of what happened is relative to the context of inquiry, to what they are seeking to establish. But the *fact* to be explained here is "relative" to the goals of inquiry, not to the belief systems which agents assent to or accept as true. Since the political historian and the physiologists have different *explananda*, what counts as evidence for the political historian will not be the same kind of evidence that will be of interest to the physiologist? But to say that the evidence adduced to provide rational support for the explanations of the physiologist and for those of the historian depends on the context of their respective inquiries is not the same as saying that justification is relative to the different belief systems that individuals such a Galileo and Bellarmine accept. It is rather to make the point that the chain of entailments that lead from the presentation of a fact as political (as an action) means that the question "why did Sergei Skripal collapse?" is not answered by invoking the kind of (nomological) explanation which cites an antecedent condition (the physical contact with the nerve agent) together with the general law that contact with nerve agents leads to death unless treated. For once something is presented in our discourse as a (political) act it commands an explanation of a different (intentional/teleological) kind.[4]

This form of explanatory pluralism is not the same relativistic pluralism that Boghossian (rightly or wrongly) attributes to Rorty and which constitutes the target of his critique in *Fear of Knowledge*.[5] Rorty's relativism holds that *rational justification is relative to the belief system an agent or a group accepts.* This claim is an inference from the following premises:

(i) There are different belief systems (an uncontroversial factual or descriptive premise).
(ii) The conditions of knowledge are not truth-evaluable.
(iii) Therefore: rational justification is relative to the belief system an agent or group accept.

Nonrelativistic pluralism, on the other hand, holds that rational justification is relative to the goals of inquiry. This claim is an inference from the following premises:

(i) Explanatory inferences should be fit for purpose.
(ii) The conditions of knowledge are not truth-evaluable.
(iii) Therefore, rational justification is relative to the goals of inquiry.

Can the hinge epistemology at work in Collingwood's account of absolute presuppositions evade the most common objections levelled against the relativistic pluralism which Boghossian takes to be the inevitable result of denying that the conditions of knowledge are truth-evaluable? The standard objections raised against epistemic relativism are that (i) it is not possible to criticize the epistemic judgments of other cultures or belief systems and (ii) that relativistic pluralism is self-undermining. I shall consider these objections in the next section.

4 Epistemic Pluralism Is Neither Quietist nor Self-Undermining

Let us rewind to Boghossian's presentation of the dispute between Cardinal Bellarmine and Galileo over the truth of heliocentrism. On Boghossian's reconstruction of Rorty, the relativist cannot say that heliocentrism is true and geocentrism is false. Since there is no fact of the matter that is independent of the belief system that Bellarmine or Galileo in each case accepts, Galileo and Bellarmine are justified in holding something as true only relative to the belief system that each one of them accepts. At best, according to the relativist, Galileo can say that heliocentrism is true and that one is justified in holding it to be true given the belief system that he, Galileo, accepts. Of course, Galileo could say "Bellarmine is wrong" but this would be nothing but shorthand for "according to the belief system that I (Galileo) accept Bellarmine is wrong." Since rational justification is relative to the belief system that agents accept, there is no way of adjudicating the dispute between geocentrism and heliocentrism. Bellarmine and Galileo simply must agree to disagree or, as Rorty says, exchange insults. The denial that truth-values can be legitimately predicated of the conditions of knowledge leads to a form of malignant pluralism which relativizes justification to the belief system agents accept, thereby undermining the possibility of disagreement with, and criticism of, views that the speaker does not accept (Boghossian 2006: 58–68).

This conclusion is not one that the hinge epistemology developed by Collingwood as discussed above is committed to. The preceding discussion of the role of philosophical analysis in making explicit the absolute presuppositions at work in forms of inquiry gives us an insight into how the hinge epistemology that is implicit in Collingwood's account of absolute presuppositions would handle cross-cultural disagreements. Consider the case of a group of Egyptologists who enlist the help of a team of forensic archaeologists to establish whether the mummified remains of a pharaoh are genetically related to those of a mummy found in another sarcophagus. Using DNA analysis, the forensic team discovers that Tutankhamun was the product of an incestuous union, and that is why he had a club foot and various genetic disorders that led to his early death. The DNA tests used by the forensic archaeologist presuppose the principle of the uniformity of nature. Just as palaeontologists who explain why dinosaurs became extinct when the Earth's temperature dropped presuppose that cold-blooded creatures uniformly thrive in similar (warm) environmental conditions, so forensic archaeologists could not establish that Tutankhamun developed a club foot as a result of inbreeding without presupposing that DNA behaves in the same way in Ancient Egypt as it does in the present day. *Qua* historians, Egyptologists presuppose that the epistemic norms by which agents lead their lives change from time to time, and they explain the incestual relations which led to the birth of Tutankhamun by invoking the epistemic norm that inbreeding keeps the blood pure. When relying on the results of DNA testing, on the other hand, historians of ancient Egypt switch, as Carnap might say, from one explanatory framework to another, and presuppose the same thing as natural scientists do, namely, that nature is uniform.[6] If asked whether they agreed with the ancient Egyptian's view that incest was good for the blood, they need not accept a quietist position for fear of being accused of anachronism in the following way: but the Egyptians did not know about DNA, therefore you cannot invoke DNA as an explanation for Tutankhamun's club foot by the Ancient Egyptians' lights! As far as the Ancient Egyptians were concerned DNA did not exist and since what exists or is true depends on what is believed to be true you cannot judge the Ancient Egyptians for what they believed to be the case! The Egyptologist who denies that incest keeps the bloodline pure switches from the presuppositions which govern the study of past cultures to the presuppositions of natural science. This switch in presuppositions is justified by the fact that the explanatory goals of the Egyptologist and those of the forensic archaeologist are not the same. Their explanations answer different kinds of questions which require not different explanations but explanations of a different kind. *Qua* historians, Egyptologists are interested in why incest was common amongst the Royal family and explain it by invoking the norm that inbreeding keeps the blood pure. *Qua* natural scientists forensic archaeologists explain why Tutankhamun had a club foot by appealing to the generalization that inbreeding leads to genetically inherited diseases. Just as the political historian's and the physiologist's account of Sergei Skripal's collapse are directed at different, inquiry-specific *explananda*, so Egyptologists and forensic archaeologists do not explain the same thing or fact. But, and this is the crucial caveat, while the facts which are explained by Egyptologists and forensic archaeologists are not the same facts (one explains why Ancient Egyptians engaged in incestual relations [to keep the bloodline pure] and the other why Tutankhamun had a club foot [because he was the product of inbreeding]), these facts are not relative either to what the Ancient Egyptians believe (that engaging in incest keeps the blood pure) or to what contemporary historians of Ancient Egypt believe to be the case (that inbreeding leads to genetic disease). They are "relative" to the explanatory frameworks of natural science and humanistically oriented anthropology. By switching to the presuppositions of natural science twentieth-century historians can say that the Ancient Egyptians were wrong and that incest leads to genetic malformations rather than ensuring the purity of the blood, without committing the anachronism of explaining the behaviour of Ancient Egyptians by invoking epistemic norms which were alien to them. Let us transpose this to the Bellarmine/Galileo dispute. When taking the perspective of science (rather than identifying with what Galileo believes to be true/accepts as true), one can say that Bellarmine was wrong when stating that the universe is geocentric, just as the twentieth-century historian would deny inbreeding keeps the blood pure. It is not necessary to assume that the presuppositions which are foundational for natural science (the principle of the uniformity of nature) are true of an inquiry-independent reality to be able to pass critical judgments on the epistemic claims of different cultures. What is crucial, on the other hand, is to disambiguate the explanatory goal of historians from those of forensic archaeologists and to take care to note what hat one is wearing when making the claim that Tutankhamun had a club foot as a result of inbreeding.

But (I hear a potential objector claim), according to the epistemic pluralist the perspective of science (with its reliance on the principle of the uniformity of nature as a condition of the possibility for providing empirical evidence) belongs to the belief system that Galileo accepted, and Bellarmine rejected. The claim that heliocentrism is justified by the inductive/scientific method with its reliance on the principle of the uniformity of nature, so the objection goes, is shorthand for "according to my post-Galilean epistemic system I am justified in believing that the universe is heliocentric". Since the perspective of science is a belief system which Galileo accepted, and Bellarmine rejected, it is not possible to avoid quietism while rejecting the view that explanatory frameworks can be evaluated for truth and falsity. This is a conclusion that some postmodern philosophers of science have been all too ready to endorse. Famously or infamously Latour (2000) claimed that since the Koch bacillus was not discovered until 1882 it is not possible to say that Ramesses II died of tuberculosis because such claims would have made no sense within the belief system of the Ancient Egyptians. Indeed the Koch bacillus did not exist at the time. This is the sort of paradoxical conclusion to which the relativistic pluralism that Boghossian criticizes leads and which Boghossian (2006: 26 and 113) is (rightly) anxious to avoid. However, while some have indeed drawn these relativistic conclusions, merely denying that the conditions of knowledge are truth-evaluable does not mandate this conclusion. If the preceding analysis of the account of absolute presuppositions is correct, this is not a conclusion that the hinge epistemology implicit in Collingwood's account of absolute presuppositions would have to accept. What makes the principle of the uniformity of nature operative in science is not that "I", the subject, accept it, endorse it, or believe it to be true, or that "we", my culture, accepts it, endorses it and largely holds it to be true, but that the principle of the uniformity of nature is entailed by the offering of a certain kind of (empirical) evidence in support for one's claims. Whether the principle of the uniformity of nature is a principle which governs inductive inferences, therefore, does not depend on whether Galileo or anyone accepts the scientific worldview, nor does the answer to the question whether the Earth revolves around sun (or the other way around) depend upon what Bellarmine or Galileo believe to be the case. To say that Galileo accepts the principle of the uniformity of nature does not add anything to the claim that the principle of the uniformity of nature is logically entailed by the making of inductive inferences. Just as *Mary's believing or accepting* that the concept of having three sides is entailed by the concept of a triangle does not do any logical work, so *Galileo's believing or accepting* that the principle of the uniformity of nature is entailed by the making of inductive inferences adds nothing whatsoever to the claim that the principle of the uniformity of nature is entailed by the making of inductive inferences. As we have seen, it is the nature of presuppositions to do their logical work not in so far as they are believed to be true or false (as if they were propositions) but in so far as they are entailed by the propositional answers given to the questions asked. It is not Galileo's acceptance of the principle of the uniformity of nature that justifies him in the making the inductive inferences which are characteristic of the belief system of science. It is rather that no one advancing empirical claims can deny the principle of the uniformity of nature at the cost of internal inconsistency. Collingwood would agree with Boghossian that we can appeal to the inductive method of natural science, with its reliance on the principle of the uniformity of nature, to correct the epistemic claims of other cultures. Boghossian, for example, argues that the belief system of Bellarmine and that of Galileo were not actually as different as they look. For ordinary propositions about medium-sized goods Bellarmine used exactly the same empirical method as Galileo and, when doing so, relied on the same principle underpinning the belief system of science. Where Bellarmine differed from Galileo was in assuming the Bible provided empirical evidence for the nature of the Earth's rotation (Boghossian 2006: 102 ff.). Thus one can correct the claim that the universe is geocentric by appealing to evidence of the same empirical kind, albeit of a more sophisticated nature. While Collingwood would agree that it is possible to use the principle of the uniformity of nature cross-culturally, he would deny that it is necessary to hold that the principle of the uniformity of nature is true of an inquiry-independent reality in order to invoke it in a cross-cultural context. For the applicability of this principle is justified (even if only in a backhanded way) by the fact that it enables certain predictive claims through induction, not by its being true of an inquiry-independent reality or by its being believed (by someone such as, e.g., Galileo) to be true. Collingwood accepts that absolute presuppositions cannot be justified in a non-circular way: since they are constitutive of the explanatory practices which they make possible, rather than being strictly universal, they derive their validity from the fact that they underpin certain explanatory practices. What he denies is that they play their logical role as enablers of knowledge in so far as they are believed by someone to be true. And this is the reason why his explanatory pluralism is ultimately not a form of relativism. For the relativist, a belief system derives its legitimacy from its being endorsed, accepted, or believed as true. Explanatory pluralism, on the other hand, matches logical/explanatory form with the explanatory goal. This enables the pluralist to assess explanations on the basis of their fitness for purpose.

Denying that the conditions of knowledge are truth-evaluable does not lead to quietism if epistemic pluralism is understood as claiming that explanation is dependent on the context and goals of inquiry and if it is decoupled from the view that the true and false claims which arise within forms of inquiry are relative to the people who accept them. One need not, therefore, commit to the view that there are inquiry-independent truths, in order to engage critically with other cultures. There is an intermediary position between the realist's adherence to the correspondence theory of truth and the extreme postmodernism which, according to Boghossian, is entailed by its rejection. Absolute presuppositions do not have any causal powers; they neither bring about the existence of the Koch bacillus nor do they have the power to make the sun revolve around the Earth. Invoking the threat of postmodernism to bolster the correspondence theory of truth is arguably a form of philosophical scaremongering.

Collingwood's explanatory pluralism is thus not vulnerable to the first standard objection that is often raised against the relativism about belief systems, namely, that it undermines the possibility of epistemic disagreement. But is it vulnerable to the second objection that is often raised against it, the charge of internal inconsistency? The charge runs as follows: the relativist claims knowledge to be relative to the belief system that the subject accepts or holds to be true. There are two ways in which such a claim can be understood. On the one hand, it can be understood as stating an absolute truth, namely, that all knowledge is relative to the belief system that the subject accepts; in this case, the position of the relativist would be incoherent because, as Boghossian says, the relativist would be committed to at least *one* absolute truth, namely, that all knowledge is relative to the belief system that the subjects accept. On the other hand, it can be understood as stating what the moral relativist believes, namely, that there are no facts which are true independently of the belief system that the subject accepts; in this case the relativist's claim can be safely ignored because it would be simply a report about what the relativist finds it agreeable to believe (Boghossian 2006: 53). Relativists are therefore caught between two horns of a dilemma: if they exempt their claim from being subject to the relativist clause (I believe/my culture believes…) they become guilty of making an exception for their own position; if they act on the second horn of the dilemma then they make their own relativist metaphilosophy contingent upon the belief system that a subject accepts.

Is the account of absolute presuppositions that lies at the core of Collingwood's hinge epistemology vulnerable to this criticism? One might think that it is. The standard self-refutation objection that is levelled against belief-system relativism could be adapted and directed against Collingwood as follows: Collingwood claims that knowledge rests on absolute presuppositions. This claim entails that there is no presuppositionless knowledge, that all knowledge rests on presuppositions. But the claim "all knowledge rests on presuppositions" is a universal claim. QED: like standard forms of belief system relativism, Collingwood's explanatory pluralism is self-refuting because it advances at least one universal knowledge claim, namely, that all knowledge rests on presuppositions.

Can Collingwood's hinge epistemology escape this charge? One might think that it cannot because the escape route out of the self-undermining objection that is open to the relativist is not available to his version of hinge epistemology. When accused of inconsistency relativists can respond to the self-undermining charge by saying that the apparently universal claim they appear to be making is not, appearances to the contrary, a universal claim because "all knowledge is relative to the belief system that a subject accepts" is shorthand for the genuinely relativist claim "I believe/my culture believes that all knowledge is relative to what the belief-system that the subject accepts." As we have seen, this reply is one that is often found unsatisfactory because if relativists are simply describing what a subject or their culture believes, then they can safely be ignored since they are not advancing an argument. Be that as it may, this is not the kind of reply that Collingwood would or indeed could offer because presuppositions (as we have been at pains to argue) do not do their logical work in virtue of being accepted as true (remember: it is not the pathologists believing or accepting as true that events have causes that can be tampered with, that makes the presupposition operative). Unlike the relativist, therefore, Collingwood would/could not deflect the self-undermining objection by suggesting that "(all) knowledge requires presuppositions" can be rewritten as "I believe/accept that knowledge requires presuppositions," thereby making his metaphilosophical stance contingent on its being accepted as/believed to be true by someone. Collingwood's metaphilosophical view (that knowledge rests on presuppositions) is not one on which he would be happy to agree to disagree with a realist, by proposing that whereas realists believe that there is presuppositionless knowledge he does not believe there is. Since the search for knowledge requires answering questions, and any questions we ask entail some presupposition or other, the notion of presuppositionless propositional knowledge is nonsensical. The appropriate response to the person who says, "I don't believe all knowledge requires presuppositions," therefore, would be *not* "you are entitled to your own belief as much as I am entitled to mine" *but* "you do not understand what knowledge is!" just as the appropriate response to someone who said, "I don't believe triangles have three sides" would be not "well everyone is entitled to their own opinion!" but "you do not understand what a triangle is!" Since the notion of presuppositionless knowledge is nonsensical the claim "there is no presuppositionless knowledge" is not the kind of claim that can be coherently believed to be true by one person and false by another, just as "triangles are three sided figures" is not the kind of claim which can be believed to be true by Jill and false by Jack. Just as it is not my believing that a triangle has three sides that makes it the case that a triangle has three sides, so it is not someone's disbelieving that a triangle has three sides that makes it false that a triangle has three sides. But if the claim "knowledge requires presuppositions" is not relative to a belief system that one can choose to accept or reject, is Collingwood making the very universal knowledge claim that prompts the self-undermining objection in the first instance? Arguably not, because there is a distinction between claiming that all knowledge is relative to the belief system that the subject accepts and that all knowledge rests on presuppositions. What generates the self-undermining objection in the case of belief-system relativism is the fact that the relativist is making an exception for her own knowledge claim. The relativist claims (i) that knowledge is relative to the belief system that the subject accepts and (ii) that her own claim "knowledge is relative to the belief system the subject accepts" is exempt from this requirement, at least if she chooses the first horn of the dilemma and does not seek to deflect the self-undermining objection by making her own (relativist) metaphilosophical stance contingent upon what someone believes. When making the claim that knowledge rests on presuppositions, on the other hand, Collingwood is not making an exception for the claim that "all knowledge rests on presuppositions", thereby committing to the view that there is at least one knowledge claim that is presuppositionless. Philosophy, unlike the special sciences, does not advance any knowledge claims: philosophical analysis uncovers what presuppositions are operative whenever a knowledge claim is advanced. Since philosophy does not advance knowledge claims, it does not have presuppositions of its own; any presuppositions philosophy deals with are borrowed from the sciences, which are the laboratories of knowledge. Through presuppositional analysis, we come to understand how knowledge in the sciences is produced, but this philosophical understanding is not a knowledge claim, like that of the historian, of the chemist, of the physicist, the mathematician and so on. The suspicion that there may be foul play and that Collingwood, like the relativist, must choose between either making his metaphilosophical stance contingent or risking internal incoherence, rests on conflating philosophical knowing with knowledge in the sciences. Philosophy, as he says,

> does not, like exact or empirical science, bring us to know things of which we were simply ignorant, but brings us to know in a different way things which we already knew in some way.... (EPM: 161)

The claim "knowledge rests on presuppositions" is not itself a knowledge claim, but an analysis or explication of what knowledge is.

The worst accusation that can be raised against a philosophy is that it fails by its own standards. It would not seem too much to ask of a philosopher that their metaphilosophical views should be compatible with their first order claims since one might expect consistency to be a paradigmatic philosophical virtue. Yet examples of such inconsistencies between a philosopher's first and second-order views are not uncommon. Take Ayer's logical positivism, for example. It appeals to the principle of verifiability to launch a critique of metaphysical propositions about transcendent metaphysical entities. But, it is often asked, what kind of proposition is the principle of verifiability? If it is neither an empirically verifiable proposition, nor can be exempted from the verifiability requirement because that would make it merely tautological, then positivism's criticism of metaphysics applies to its own metaphilosophical commitment to the view that propositions which are neither empirically verifiable nor mere tautologies are meaningless. The kind of hinge epistemology at work in Collingwood's account of absolute presuppositions is not vulnerable to the charge of self-refutation.[7]

5 Conclusion

Let us recap what this chapter has and has not established. First of all, it was not my purpose to establish whether Boghossian's interpretation of the belief system relativism that he attributes to Rorty and Wittgenstein is correct. Nor was the goal of this chapter either to assess or undermine Boghossian's critique of belief system relativism. The goal of this chapter was rather to decouple belief system relativism from explanatory pluralism by challenging the assumption that rejecting the claim that the conditions of knowledge have truth-makers inevitably leads to the sort of postmodern relativism Boghossian is keen to undermine. I have done so by presenting a reading of Collingwood's account of absolute presuppositions as epistemic hinges which are entailed by the knowledge claims made in different contexts of inquiry, and by arguing that the validity of these hinges is not relative to their being accepted or believed to be true by individuals or cultures; such hinges are rather justified (in a backhanded way) by the kinds of explanations which they undergird. While presuppositions such as "Earth is flat" or "the universe is geocentric" are relative to the explanatory framework of natural science and can be dislodged by the scientific method, some presuppositions, such as the principle of the uniformity of nature, cannot be dislodged in that way because they can never be proved or disproved to be true or false inductively. The distinction between relative and absolute presuppositions supports the claim that there are "propositions" which have a special logical status and which lie beyond verification, not in the logical positivist's sense that if subjected to this test they would fail it, but in the sense that one cannot conceive of a possible scenario in which they may be falsified without undermining our ability to reason. Understood in this way absolute presuppositions capture Coliva's recent interpretation of Wittgenstein's hinge propositions in *On Certainty* as supplying the conditions of epistemic rationality (Coliva 2015: 157 ff.). Just as Coliva's reading of Wittgenstein's hinge epistemology highlights certain aspects of the text at the expense of others, so my presentation of

Collingwood's account of absolute presuppositions has downplayed the historicist passages in *An Essay on Metaphysics* which have informed the standard reading of Collingwood's metaphysics (Toulmin 1972; Rotenstreich 1972; Donagan 1962), a reading which would bring Collingwood's epistemic pluralism closer to the belief system relativism that Boghossian ascribes to Rorty, but which would also make it vulnerable to the twin charges of quietism and internal incoherence that belief-system relativism struggles to answer. There are always, inevitably, hermeneutic choices to be made. I have made mine fully cognizant of the fact that the interpretative choices one makes reflect the kind of philosopher one is.

Notes

1. Hinge epistemology is often discussed in the context of Wittgenstein.
2. In this example of a relative presupposition Collingwood is referring to a measuring tape, not to the unit of measurement, the foot. The question "is the foot accurate as a unit of measurement?" is the kind of question which Collingwood would have most likely treated as one which cannot be answered by stating either "yes the foot is an accurate unit of measurement", or "no, the foot is not an accurate unit of measurement". In other words, Collingwood would have treated the proposition "the foot is an accurate unit of measurement" not as an ordinary proposition that can be found to be true or false, but as a presupposition to which the notion of truth and falsity does not apply, and moreover as the kind of presupposition which cannot be restated as a proposition by changing its role.
3. On what is unfortunately a standard reading of Collingwood, this rejection of metaphysics allegedly leads to the dissolution of philosophy into a form of cultural anthropology. Since the task of the metaphysician is no longer to assess absolute presuppositions for truth or falsity and to discuss, for example, whether there are causes, or whether there is freedom of the will, the task of metaphysics is downgraded to that of establishing what people tended to believe at certain times and places. On this view metaphysics simply gets dissolved into history. While there are passages that lend support to the standard interpretation of what philosophy is supposed to do (to discuss the truth or falsity of the most general propositions about reality), the standard interpretation is clearly at odds with Collingwood's view that metaphysics (understood as the study of absolute presuppositions) is a logical enquiry and that the philosopher is a kind of detective who uncovers presuppositions by following the logical clues that they leave in the questions to which they give rise. Thus while the explanatory practices that meta-

physics studies are historically instantiated, the logical regress from answers to questions, and from questions to presuppositions, is not a descriptive historical process. The standard interpretation of *An Essay on Metaphysics* as dissolving metaphysics into history brings Collingwood closer to the kind of relativistic pluralism of Rorty. But such an interpretation fails to acknowledge the full implications of Collingwood's claim that presuppositions do not possess causal power, but logical efficacy, and therefore that they stand to the questions to which they give rise as their logical rather than causal ground.

4. Collingwood was interested in the form of inferences which govern the explanation of action and of events. These received extensive treatment by Dray (1963, 1967, 1980).
5. My goal is not to evaluate whether Boghossian's reconstruction of Rorty is correct.
6. For a comparison of Collingwood and Carnap on this see D'Oro 2015.
7. There is a third objection against epistemic relativism which Boghossian deems to be even more powerful than the self-undermining objection. This is the claim that epistemic relativism advances incomplete propositions (Boghossian 2006: 87 ff.) which potentially generate a vicious regress (2006: 89). As Boghossian says, it is crucial to the relativist that a thinker *accepts* a belief system (Boghossian 2006: 86). This entails that propositions such as "Copernicanism is justified by Galileo's observations" is incomplete and stands for "In relation to epistemic system C (which the subject accepts), Copernicanism is justified by Galileo's observations". I have not considered this objection in much detail because, although distinct from the self-undermining objection, it relies on a premise to which the hinge epistemology that I have ascribed to Collingwood is not wedded: Collingwood, as we have seen, denies that presuppositions do their work in virtue of being believed or accepted as true by the subject.

Bibliography

Boghossian, P. 2006. *Fear of Knowledge*. Oxford: Oxford University Press.

Carter, A.J. 2017. Epistemic Pluralism, Epistemic Relativism, and 'Hinge' Epistemology. In *Epistemic Pluralism: Methodological Issues*, ed. Annalisa Coliva and Nikolaj Jang Lee Linding Pedersen, 229–249. London: Palgrave Macmillan.

Coliva, A. 2010. Was Wittgenstein an Epistemic Relativist? *Philosophical Investigations* 33 (1): 1–23.

———. 2015. *Extended Rationality: A Hinge Epistemology*. London: Palgrave Macmillan.

Collingwood, R.G. 1933. *An Essay on Philosophical Method*. Revised with an Introduction by James Connelly and Giuseppina D'Oro. Oxford: Oxford University Press, 2005.

———. 1940. *An Essay on Metaphysics*. Revised Edition with an Introduction by R. Martin. Oxford: Oxford University Press, 1998.

D'Oro, G. 2002. *Collingwood and the Metaphysics of Experience*. London/New York: Routledge.

———. 2010. The Myth of Collingwood's Historicism. *Inquiry* 53 (6): 627–641.

———. 2015. Unlikely Bedfellows? Collingwood, Carnap and the Internal/External Distinction. *British Journal for the History of Philosophy* 23 (4): 802–817.

Donagan, A. 1962. *The Later Philosophy of R. G. Collingwood*. Oxford: Clarendon Press.

Dray, W.H. 1963. The Historical Explanation of Actions Reconsidered. In *Philosophy and History*, ed. S. Hook. New York: New York University Press.

———. 1967. *Laws and Explanation in History*. London: Oxford University Press.

———. 1980. R. G. Collingwood and the Understanding of Actions in History. In *Perspectives on History*, 9–26. London: Routledge and Kegan Paul.

Festin, R. 2009. Collingwood and Wittgenstein on Magic. *Collingwood and British Idealism Studies* 15 (1): 41–70.

Hamilton, A. 2014. *Routledge Philosophy Guidebook to Wittgenstein and on Certainty*. Abingdon: Routledge.

Latour, B. 2000. On the Partial Existence of Existing and Non-Existing Objects. In *Biographies of Scientific Objects*, ed. L. Daston, 247–269. Chicago: University of Chicago Press.

Moyal-Sharrock, D. 2004. *Understanding Wittgenstein's on Certainty*. London: Palgrave Macmillan.

Moyal-Sharrock, D., and William H. Brenner, eds. 2007. *Readings of Wittgenstein's on Certainty*. Basingstoke/New York/London: Palgrave Macmillan.

Rorty, R. 1981. *Philosophy and the Mirror of Nature*. Princeton: Princeton University Press.

Rotenstreich, N. 1972. Metaphysics and Historicism. In *Critical Essays on the Philosophy of R. G. Collingwood*, ed. M. Krausz, 179–200. Oxford: Clarendon Press.

Toulmin, S. 1972. Conceptual Change and the Problem of Relativity. In *Critical Essays on the Philosophy of R. G. Collingwood*, ed. M. Krausz, 201–221. Oxford: Clarendon Press.

Wittgenstein, L. 1969. *On Certainty*. Malden: Blackwell.

CHAPTER 8

Oscillation and Emancipation: Collingwood on History and Human Nature

Karim Dharamsi

1 Introduction

To what extent we are creatures endowed with 'conceptual sovereignty', *free* and answerable to legitimately *sui generis* orders of justification, depends, at least in part, on what relationship we think obtains between reason and nature (McDowell 1994: 134; Quine 2013: 87). Philosophical naturalists assess philosophy's efficacy by how well it aligns methodologically with the empirical sciences. They believe that philosophy should follow the methods of the natural sciences, thereby calling into question the central tasks of traditional epistemology, namely, the study of the conditions of justification. Philosophical naturalists are not merely recommending a casual transformation. This recommendation, underwritten by explicit ontological commitments, is far-reaching. It enjoins a rethinking of how we conceive of philosophy's autonomy and so its methodologies,

This chapter owes much to Stephen Leach's and Giuseppina D'Oro's thoughtful comments and recommendations.

K. Dharamsi (✉)
Mount Royal University, Calgary, AB, Canada
e-mail: kdharamsi@mtroyal.ca

K. Dharamsi et al. (eds.), *Collingwood on Philosophical Methodology*, Philosophers in Depth,
https://doi.org/10.1007/978-3-030-02432-1_8

and it may also impact upon how we understand human nature and our place in the order of things.

R.G. Collingwood's philosophy of historiography and what Giuseppina D'Oro (2003) calls 'anti-naturalism' may seem, at least on first blush, to align well with positions defended by some liberal naturalists, those committed to the methodological autonomy of philosophy while accepting some kind of ontological monism (EM: 50–51; D'Oro 2003: 53; McDowell 1994: 108). After all, Collingwood is a monist, even though he is not a physicalist. In *The Principles of History*, Collingwood writes that "[h]istorical thought, thought about rational activity, is free from the domination of natural science, and rational activity is free from the domination of nature" (PH: 101). Here, Collingwood is making two substantive points. He defends a form of methodological pluralism, such that the methods for the study of human action are distinct from the methods of the natural sciences. Furthermore, he distinguishes the *sui generis* character of rational activity from determinate nature. Later in the same text Collingwood admonishes attempts at a 'science of human nature', which he believes endeavours to treat human reason 'non-criteriologically', "that is, without reference to the self-critical element"—the hallmark of justificatory norms. Collingwood takes the criteriological 'element' as being the essence of rational activity, insisting that any 'science of reason' "which ignores it is not a science of reason and throws no light on its nature" (PH: 108). In other words, in ignoring the normative conditions of justification, *the essence of rational activity*, a science of reason is uninformative.

Some have credited John McDowell's liberal naturalism (1994) with defending an attractive alternative to mainstream philosophical naturalism (De Caro and Macarthur 2010: 2; Macarthur 2018). Against various versions of the myth of the given, he argues that our conceptual capacities are at work not only in obvious cases of inferential reasoning but also in perceptive experience (Aportone 2011: 75; Smith 2002: 61). Exploiting Kantian language, McDowell is interested in establishing a link between sensibility and understanding, and between receptivity and spontaneity. He dismisses the views of naturalists who believe that rational norms are reducible to some kind of law-governed explanatory model and is equally suspect of dualisms of scheme and content. In *Mind and World* McDowell seems to accede to an idealism that coincides with the realism of common sense. He thereby attempts to plot what may be a middle path between two extremes of an 'oscillation'—bald naturalism, on the one side, and a conceptual coherentism that appears entirely amputated from the empirical world, on the other (Aportone 2011: 77; McDowell 1994: 26; Pradhan 2015: 1–2). He sometimes refers to his middle path as a form of *naturalized Platonism* (McDowell 1994: 27).

Where Collingwood is suspicious of any attempts at a 'science of human nature' that ignores the constitutive character of rationality and the centrality of a historical consciousness, McDowell is suspicious of any account of human nature that fails to provide for the immediacy by which our concepts 'latch onto' the world and a maturational process by which we fully achieve the *sui generis* character of rational personhood. It is by way of this maturational *becoming*, of acquiring a 'second nature', that McDowell insists we attain a *gapless* relationship between mind and world. By saying that there is "no gap between thought, as such, and the world" is simply, according to McDowell, to "dress up a *truism* in high-flown language" (McDowell 1994: 27). In a number of places in *Mind and World* he reveals that an identity theory of truth is core to his rethinking of mind-world relations and he takes such a theory as expressing a truism (Dodd 1995: 161).

A perfunctory assessment may suggest a significant overlap between Collingwood and McDowell since both have stressed arguments for philosophy's autonomy and against the kind of reduction naturalists have tended to support. Indeed, as early as his 1922 paper, "Are History and Science Different Kinds of Knowledge?" and in his 1924 book, *Speculum Mentis*, Collingwood disavows givenness, or the idea of non-conceptual contents playing a justificatory role in empirical judgement (SM: 448; Collingwood 1922: 205). A critique and rejection of givenness is vital to McDowell's philosophy of mind. Both Collingwood and McDowell are interested in protecting human freedom and rational norms against the determinacy of nature, and there does appear to be a shared commitment to history.

This list implies notable similarities. However, in this chapter, I examine two significant and related differences that, if I am right, would imply that Collingwood and McDowell present different views of mind-world relations that rest on competing views of human nature.

First, I take some time to set up the discussion around naturalism and Quine's naturalization of epistemology. It is Quine's naturalism that figures as a backdrop to McDowell's argument. I offer some general Collingwoodian responses to this backdrop. I then, second, examine McDowell's *liberal* recasting of naturalism. I suggest that McDowell's naturalism rests on his maturational account of personhood. Finally, I argue that Collingwood's historically oriented metaphysics and metaphilosophical considerations are at odds with McDowell's positive claims. While they may share superficial structural similarities in their attacks on naturalism, McDowell's liberal naturalism is significantly different from Collingwood's.

A key to Collingwood's defence of the irreducibility of reasons and norms is the idea that human nature is not specifiable once and for all. He combines this *transcendental commitment* with a procedural view of persons as *distinctively* historical beings (PH: 193–94).[1] While McDowell is committed to a normative account of persons, and he does reference tradition and history, it is clear that his approach is not *historical* in Collingwood's sense.

2 Naturalism and Method

In *An Essay on Philosophical Method*, Collingwood assesses the status of philosophical inquiry in the nineteenth century (EPM: 5). Having provided a brief accounting of the appropriation of Cartesian philosophy in science and Kant's philosophy in humanistic study, Collingwood writes, "[t]hroughout the greater part of the nineteenth century, the attention of active minds was chiefly taken up by these two branches of knowledge, science and history; there seemed nothing left for philosophy to do, and it sank into complete neglect, except as an appendage of natural science or as a part of history" (EPM: 5). While Collingwood is remarkably optimistic about the future of the discipline, noting fresh ideas and "a level of philosophical work which no one could have anticipated fifty years ago [at the end of the nineteenth century]" he does offer a caution: "… in order that this promise should be redeemed, one thing is needed above all others: a patient and thorough reconsideration of the problem of method" (IH: 5). In a fashion, Collingwood may have been prescient, at least in cautioning philosophers against the possibility of methodological complacency. Subsequently, for some philosophers, at least those committed to strictly physicalist renderings of naturalism, questions of method seem to be settled. We might think, if disposed along these lines, that we are now in the age of fine-tuning, asking what counts as natural and what does not—which may be tantamount to admitting that we are not entirely clear on what we think naturalism really is—and are unhappy with the cost of making too narrow or too broad a commitment to its scope depending on what relationship between reason and nature we are interested in rejecting or defending.

'Naturalism' is notoriously difficult to define but is arguably the defining orientation of philosophical methodology today. 'Naturalization' is what those committed to 'naturalism' do. Collingwood observes that philosophers are unreflectively "under the domination of methodological ideas inherited from the nineteenth century, when philosophy was in various ways assimilated to the pattern of empirical science" (EPM: 7). This trend has continued. According to Robert Audi's (2011): "[c]ontemporary philosophy is witnessing an age of naturalization projects: in ethics, in philosophy of mind, and even in philosophy of mathematics and logic" (Audi 2011: 169). These projects share in common an influence of the nineteenth-century thinking on the disposition of philosophical work today and the attention to its recasting canonical questions in metaphysics and epistemology in the key of science (Ibid: 169).

Of course, naturalism is not new and can also include the philosophy of history on Audi's list.[2] Various forms of naturalism have defenders and detractors who have been around for as long as philosophical inquiry itself.[3] Notwithstanding some elements common to all its forms or ordinary intuitions informing *contemporary* uses of 'naturalism', what requires 'naturalization' unsurprisingly resists consensus amongst philosophers. Michael Rea (2002) in his *World Without Design: The Ontological Consequences of Naturalism*, jests that "[o]ne who believes with Parmenides and Plato that a priori reflection is a more secure way of knowing than empirical investigation can well be assured of a cold reception in the house of naturalism" (Rea 2002: 22). Rea thinks that beyond this there is little agreement since naturalism "is identified sometimes with materialism, sometimes with empiricism, and sometimes with scientism" (Ibid: 22). These '-isms' are related, with the last perhaps representing the most muscular fidelity of philosophy to science—and reflecting a high opinion of the contribution of science to human flourishing. Still, Rea is right to observe that these positions "are equally difficult to characterize and, in any case, the identifications are controversial" (Ibid: 22). Even when we grant naturalists a commitment to nothing but nature and accept that there is nothing *supernatural*, there is "widespread disagreement among naturalists about what it is for something to count as natural or supernatural" (Ibid: 22). Audi offers some sobering *ontological* grounds for agreement amongst naturalists thereby reinforcing that no matter the disputes about the scope of naturalism, it is an ontological thesis that invites methodological responsibilities:

> There is no generally accepted definition of 'naturalism', but philosophical naturalists agree that the 'descriptive' concepts and properties that must be countenanced to account for inquiry in the natural sciences should count as natural. What are often called 'observation concepts', paradigmatically those whose application is determinable through using five senses, are clear and basic cases of the descriptive. With all this in mind, we might consider naturalism, in its philosophical as opposed to its methodological form, to be roughly the view that nature is all there is and the only basic truths are the truths of nature. Supernatural entities are thus ruled out, but only examination of specific versions of naturalism will tell us whether, for instance, mental phenomena must, on those views, be reduced to physical ones to count as natural. (Audi 2011: 170)

In contrasting the 'philosophical' with the 'methodological' Audi takes the former to characterize our ontological commitments, that *nature is all there is*, and the latter our epistemological commitments that *truth* involves some kind of alignment between the application of our concepts and the order we intend for those concepts to describe. On Audi's view, then, once we have accepted the superiority of science in describing and explaining nature, perhaps relying on it (broadly) to settle our ontological questions, we can set out to prescribe our epistemic responsibilities. Hence on a *realist* account, we can draw a straight line from the *extensionality* of our concepts to the *physics* of nature, allowing the philosophical naturalist to exclude the *supernatural* on grounds that observation concepts cannot apply to supersensible entities. Since the naturalist's underlying presuppositions give licence to exclude non-extensional entities, only sensory concepts and properties working to *describe* sensory input can be considered natural in the relevant sense (Ibid: 170). This means that mental phenomena, by Audi's lights, are adjudicated for their ontological status against the methods that seek, ultimately, to draw an extensional line between concept and world. The naturalist's bequeathing of priority to an ontological monism of the kind described seems to then demand a methodological orientation of philosophy that answers to science and the demands of empirical investigation—to which science, too, is forced to answer (Ibid: 170).

However, Rea's misgivings notwithstanding, at a purely methodological level, naturalism can be straightforwardly defined as the belief that philosophy should follow the methods of natural science. As such, chief among the strengths of naturalism is the possibility of a complete picture it can frame, extending the possibilities of explanatory closure promised of causal explanation in nature to the *apparent* recalcitrance of *sui generis* rationality. Its metaphysical presuppositions treat science as offering the most reliable methodology for any investigation into reality, and *reality* includes the fully *naturalized* workings of mind and human conduct (Quine 1981: 72).

3 Quine's Naturalized Epistemology

It is in light of this promise of a complete picture that Quine (1969) famously renounces normative epistemology. Rather than establishing philosophical foundations for conditions of justification, he defends a naturalistic picture that takes epistemology to be 'a chapter of psychology', coextensive with the study of natural phenomena—and the 'physical human subject' (Quine 1969: 82, 126–127). In 'Epistemology Naturalized', Quine (1969) moves to justify "our knowledge of truths of nature in sensory terms" (Ibid: 71). He thinks the task of philosophy is descriptive; it is neither concerned with epistemic permissibility nor an a priori propaedeutic for first philosophy (Quine 1969: 126–27). As Quine puts it:

> I see philosophy and science as in the same boat—a boat which, to revert to Neurath's figure as I often do, we can rebuild only at sea while staying afloat in it. There is no external vantage point, no first philosophy. (Quine 1969: 126–27)

While Quine is right to be suspicious of an 'external vantage point', presumably one that treats reason as entirely impartial to the objects of its study, his fondness for empirical psychology folds philosophical inquiry into the study of external processes. Instead of reflective awareness, on Quine's account, explaining how one has come to hold a belief is to establish under what conditions a belief is *caused*. Philosophy, according to Quine, is on continuum with science though tending to operate at a greater level of generality. Along these lines, Quine is committed to a form of *ontological* as well as methodological continuity between science and philosophy. Not a rejection of normativity itself, Quine's naturalization aims to weaken it, reframing its epistemic responsibilities in terms of predictive efficacy. 'For me', Quine writes, "normative epistemology is a branch of engineering" (Quine and Hahn 1986: 664–65). He replaces 'truth-seeking' with 'prediction', arguing that "the normative here, as elsewhere in engineering, becomes descriptive" (Quine and Hahn 1986: 664–65). As for human beings, we can be thought to have a form of 'conceptual sovereignty' insofar as we are creatures with beliefs, but sense experience, *not reason*, is tribunal over our justifications. To turn a well-worn phrase from Donald Davidson (2001) on its head, on Quine's account, *causes are reasons.*

Quine himself does express an explicit commitment to norms that he believes are judged against a standard; but that standard is not rationally self-generating and then applied through the internal rationalizations of a self-regulated will. As Quine puts the matter, following his strictly defined empiricism:

> I am a physical object sitting in a physical world. Some of the forces of this physical world impinge on my surface. Light rays strike my retinas; molecules bombard my eardrums and fingertips. I strike back, emanating concentric airwaves. These waves take the form of a torrent of discourse about tables, people, molecules, light rays, retinas, prime numbers, infinite classes, joy and sorrow, good and evil. (Quine 1957: 228)

The 'torrent of discourse' is assessed much like one might judge the capacity for a radio receiver to pick up a station's frequency. We sometimes have to accept that our receiver is not very good or that a corrupted signal is affecting the quality of audio. We have an idea that calibrating our receiver's mechanisms or, indeed, upgrading the radio altogether, can increase the quality of output. It is in this way that Quine emphasizes again that the 'normative is naturalized' but not 'dropped'. In an argument offered up much later, Quine (1993) writes:

> Naturalism not only consigns the question of reality to science; it does the same for normative epistemology. The normative is naturalized, not dropped. The crowning normative principle of naturalized epistemology is nothing less than empiricism itself; for empiricism is both a rule of scientific method and a scientific discovery. It is natural science that tells us that our information about the world comes only through impacts on our sensory surfaces. And it is conspicuously normative, counseling us to mistrust soothsayers and telepathists. (Quine 1993: 229)

Science is normative; implicit in its investigations and its methods are standards directing us to better calibrate our instruments, fine-tuning our empirically guided methods. When Quine uses the term 'world' here he means reality as we experience it—perhaps as a producer of signals. When our radio receiver requires fixing or upgrading we are attempting to better get at 'the world' in this narrow sense. And so in this very specific way science is normative, *offering us counsel* against 'soothsayers' and 'telepathists'. If we were to take Quine's advice, we should presumably be wary of those who attempt to explain human action in the absence of evidence from empirical psychology. As Quine puts the matter, "[the epistemologist] no longer dreams of a first philosophy, firmer than science, on which science can be based; he is out to defend science from within, against its self-doubts" (Quine 1990: 474). In the case of 'mind', the task falls to empirical psychology (Ibid: 474).

In *Essay on Metaphysics*, in *The New Leviathan* and in other places Collingwood offers criticisms of psychology, but we can interpret his criticism as capturing a conceptual problem that Quine thinks he has eliminated by naturalizing the evaluative character of rational norms. In this spirit, Collingwood's criticisms rebuke the move to subsume philosophy under the umbrella of psychology—or, indeed, of a science of mind.[4] Collingwood's criticisms of psychology are consistent with his defence of philosophy's autonomy and the criteriological element of normative rationality that seems resistant to a purely descriptive science of mind. An instructive passage comes from *An Autobiography* (1939). Collingwood writes of psychology that:

> It had been deliberately created, as any might guess who knew enough Greek to understand its name, in order to study that which is neither mind in the proper traditional sense (consciousness, reason, will) nor yet body, but ψυχή, or such functions as sensations and appetite. It marched on the one hand with physiology, and on the other with the sciences of mind proper, logic and ethics, the sciences of reason and will. And it showed no desire to encroach on its neighbours' territories until, early in the nineteenth century, the dogma got about that reason and will were only concretions of sense and appetite. If that was so, it followed that logic and ethics could disappear, and that their functions could be taken over by psychology. For there was no such thing as 'mind'; what had been so called was only 'psyche'.
>
> That is what underlies the modern pretence that psychology can deal with what once were called problems of logic and ethics, and the modern

> claim of psychology to be a science of mind. People who make or admit that claim ought to know what it implies. It implies the systematic abolition of all those distinctions which, being valid for reason and will but not for sensation and appetite, constitute the special subject-matter of logic and ethics: distinctions like that between truth and error, knowledge and ignorance, science and sophistry, right and wrong, good and bad, expedient and inexpedient. (A: 94)

Here, Collingwood claims that *psychology* has a long history and that it was created to study sensations and the appetites, and not, in any philosophically valuable way, consciousness, reason, or will. He thinks that by the end of the nineteenth-century reason and will were incorporated into the study of sensations and appetites, rendering precarity to philosophy's study of logic and ethics. It is not really the accuracy of Collingwood's history that matters here, but his assessment of our study of mind and the methods we exploit in its study. The methods shape the object and treat 'mind' as a phenomenon that extends out of determinate nature without affirming a character that is distinctive—or free. Hence Collingwood fears the eventual disappearance of the study of logic and ethics altogether. Perhaps he is worried about a philosophy of mind that favours survey responses to trolley problems and morality instincts—an experimental turn that remodels or renovates familiar philosophical practices to conform to the experimental sciences and *branches of empirical psychology*.

A recent spate of publications by *experimental* philosophers may stand as evidence of philosophers no longer longing for the so-called armchair, exploiting instead the statistical and qualitative methodologies of the social sciences (Alexander 2012; Appiah 2014; Knobe and Nichols 2013; Murphy 2014). With appeals to experimental results and the assessment of 'underlying psychological processes' there seems little explanatory interrogation of any second-order metaphysical or religious presuppositions infecting the data field (2013). Rather, *uncontroversial* common sense that assumes a naturalistic metaphysics discounts metaphysical, religious and, indeed, rational explanation (i.e., explanation by reason). Appeals to explanations in terms of causes, either psychological or neurological, are often taken as the obvious route of inquiry and the basis of *legitimate* philosophical method.

Alan Donagan (1962) takes Collingwood to be attacking psychology and in some obvious ways, he is (Donagan 1962: 157). However, Collingwood is worried about the extending of quasi-scientific methodologies to our study of mind and, by extension, meanings and norms upon which reasonable actors depend for assessing those aspects of human life untouched by the scientific orientation defended by Quine. Not every case of judgement is akin to a problem in engineering or a related challenge involving predictable outcomes. In The *Principles of History* Collingwood identifies logic and ethics as 'criteriological sciences' wherein the interest is in addressing questions of how "people distinguish successes and failures in the respective fields of theory and practice" (PH: 84). Naturalistic sciences of human reason 'superseded' these criteriological sciences. Collingwood thinks that this is a fatal error and offers some clarifications about his attitude towards 'psychology':

> I do not intend any disrespect to modern psychology in so far as it has continued and developed very successfully the work of the old psychology, the rightful owner of the name, which was originally intended to identify its field as the 'psychic' part of man's nature or activities, as distinct from his 'bodily' structure and activities on the one hand and his 'rational' activities on the other. (Ibid: 84)

A criteriological science, such as history, presents a challenge to methodological unity since the object of its study resists 'naturalization' and so cannot be understood as methodologically contiguous with the natural sciences. Collingwood's discussion of the historian's task as she or he attempts to make sense of the past is instructive.

In The *Idea of History*, Collingwood (1993) discusses the task of the historian (IH: 236). In defence of the 'autonomy of historical thought' Collingwood provides an account of the historian's craft, assigning it a relationship close to that of a landscape painter. The historian "may fancy that he is reproducing in his own medium the actual shapes and colours of natural things; but however hard he tries to do this he is always selecting, simplifying, schematizing, leaving out what he thinks unimportant and putting in what he regards as essential" (Ibid: 236). The appreciable judgements being rendered of the painting's character, its ultimate relationship to the landscape it purports to represent is the responsibility of the painter—*not of nature*. In this respect, like the painter, the historian is sovereign over his work, 'he is his own master' and 'his thought is to that extent autonomous' (Ibid: 237). The kind of autonomy that Collingwood respects betrays any kind of predictable model we might wish to apply to either the painter or the historian. Imagine, for instance, sticking with our landscape painter, if by prediction we mean simple induction. We would be challenged to apply from the one painting any kind of generalization about what future paintings might depict—even of the same landscape. Indeed, we might even question whether the painter will ever paint again![5]

Were reasons merely concretions of sense and appetite it would make sense that a kind of naturalized normativity would best be able to assess the instrumental benefits of one response over another. And we could possibly have firmer grounds for predicting outcomes, following along the patterns of past results. But when Collingwood refers specifically to the special subject matter of logic and ethics he is referring to the appreciable judgements we apply against standards of correctness that are not 'external' but 'internal' to reason itself and to the autonomy we associate with the individual agent. There is a legitimate worry that a naturalized epistemology will elide the explanatory and justificatory roles of reason and rational norms. Truth and error, knowledge and ignorance, science and sophistry, and so on are rational in character and require that criteriological aspect of judgement Collingwood defends in *The Principles of History* (PH: 101). Collingwood writes:

> Psychology is an attempt to understand man by the same methods by which modern man understood Nature. These methods, as applied to the study and consequent control of Nature, were built on the double assumption that man, the knower and controller, is intelligent; nature, the known and controlled, unintelligent: mere mechanism, blind force. When these same methods ware turned upon man, they preserve their character unchanged. They therefore assume that human nature, as the object upon which they are exercised, is unintelligent. The result is that intelligence itself is converted to unintelligence. For mind, we are given mechanism which merely enjoy the honorary title of mental; for activity we are given passive reaction to stimulus; for thought, ideas associated automatically in accordance with fixed laws. (PH: 175)

This means that, contra Quine, Collingwood is interested in the justificatory norms that qualify our reasons for this claim over that and not on the 'external' conditions or processes that describe either prudential or other quasi-epistemic conditions for our maintaining this claim over that. Perhaps more importantly, the very idea of treating 'blind nature' as providing methods for the understanding of our own character seems to assume that 'intelligence' and 'reason' are irrelevant to the study of human nature. The criteriological element remains for Collingwood of a piece with what it means to engage in the practice of reasoning—something that rational agents do and not something scientists or experimental philosophers describe. A study of that 'doing' demands different methods and so the animating worry is epistemological and not ontological. Collingwood's concern with the *sui generis* constitutive character of our logical and ethical judgments and the logical-epistemic relations between our beliefs and our actions differentiates his philosophical methods from that of Quine and other philosophical naturalists.

4 Idealized Naturalism

McDowell shares Collingwood's worries about a naturalism that treats mind and reason as methodologically homologous to our study of nature—or what the former calls the 'realm of law'. The realm of law, something that we describe and assess for its patterns, is 'governed' by causal relations, invites inductive generalizations, and may seduce us into thinking that its models of predictable outcomes of *events* can apply to human *actions*. While it is clear that McDowell accepts the underlying metaphysical presupposition informing philosophical naturalism, namely, that *nature is all there is*, he defends *spontaneity* and rational norms against a kind of reductionism that would amount to describing human action in terms of law-governed rules. His argument depends on an account of human nature, which is consistent with his realist account that takes our nature to be procedurally fixed by our biology.

McDowell's *Mind and World* begins with Sellars's attack on givenness. In 'Empiricism and the Philosophy of Mind', Sellars (1997) argues that it makes no sense to think of empirical knowledge as having a 'foundation' *if* by such a foundation we mean non-conceptual or non-epistemic facts, "whether phenomenal or behavioural, public or private" (Sellars 1997: §5). He memorably asserts that, "empirical knowledge, like its sophisticated extension, science, is rational, not because it has a *foundation* but because it is a self-correcting enterprise which can put any claim in jeopardy, though not *all* at once" (Sellars 1997: §38). Sellars is primarily attacking classical empiricism and the idea that non-epistemic sensory inputs play a justificatory role in judgement. Hence Sellars's 'self-correcting' enterprise is for McDowell connected by way of perceptive experience to a tribunal by which "experience enables the layout of reality itself to exert rational influence on what a subject thinks" (McDowell 1994: 26). Experience on this account must be conceptually articulated and not *merely* connected by a causal non-epistemic link. McDowell thinks that Quine presents an unacceptable and only a "half-hearted attempt to picture worldviews as products of rational interaction between spontaneity and receptivity" (Ibid: 141). On McDowell's reading, Quine's naturalism fails to protect rational norms; conceptual articulation is, *in every case*, an exercise of *freedom*, playing an evidentiary or, indeed, an exculpatory role in a larger self-correcting enterprise that 'enables' the layout of reality. Wary of Quine, McDowell writes, "if experience plays only a causal role in the formation of a world-view, not a justificatory role, then it does not serve as evidence at all" (Ibid 1994: 134). Both McDowell and Sellars reject the idea that there is a justificatory role for non-conceptual content. Hence both maintain that *experience is not belief* (Sellars 1997: 13).

Having accepted Sellars's critique of givenness, McDowell thinks that we are at an impasse. If we side with Quine and other philosophical naturalists we displace mind, rendering it entirely unfree and governed by the causal determinacy of brute nature. Quine's naturalism forces us to confront the threat of extricating spontaneity and freedom from the world, rendering ourselves passive, in the relevantly rational sense, to the world's determinate forces. If, on the other hand, we decide *strictly* against Quine and other naturalists, McDowell thinks "we are drawn to a coherentism that cannot make sense of the bearing of thought on objective reality" (McDowell 1994: 23). He worries that by the latter conception of mind we threaten to denigrate what we think by empirical thinking. McDowell calls the latter view 'rampant Platonism', the idea that mind is "constituted independently of anything specifically human"—detached from the world and disembodied (Ibid 1994: 66, 77). His talk of 'objective reality' again confirms McDowell's realism.

McDowell recognizes that being human involves living *in* the world and practically engaging with it. He thinks that a coherentist or 'rampant Platonist' who cannot account for contact with empirical reality is defending some sort of *unnatural* conception of our practical life *in the world*. Given that we are a species of animals *belonging to nature*, we must in some way that is consistent with the development of our *nature*, 'be able to latch on to this inhuman structure', that is the world independent of our will (Ibid: 77). McDowell argues that our biological nature, a first condition towards our maturation, is ameliorated by the realization of a 'second nature' that through a process of upbringing can render us fully *cultured* and civilized participants in community and which by way of a modest empiricism latches our conceptual capacities to 'objective reality'.

McDowell takes Kant's well-known dictum as guide: "thoughts without content are empty, intuitions without concepts are blind" (Kant 1998: B 75). Experience is not belief, but thoughts in the absence of *experiential* content are empty. Here McDowell means 'representational content'. Thoughts need to be *about* something—they need to refer and the *world* provides the 'bits of experiential intake' that makes reference possible (McDowell 1994: 4). He imagines thoughts without content, and so without connection to intuitions, to not be thoughts at all.

McDowell's neo-Kantian framework takes its cue from Sellars's (1979) conceptual dualism. The former differentiates between freedom and determinacy using Kant's distinction between spontaneity and receptivity, but he exploits Sellars's understanding of the faculty of receptivity as wholly passive and noncognitive. Sellars describes by way of a neologism in *Naturalism and Ontology* a dual character to the explanatory efficacy of language, split by the orders to which each belongs and revealed by "the Janus-faced character of languagings as belonging to both the causal order and the order of reasons" (Sellars 1979: V, §64). Sellars writes that "[t]his way of looking at conceptual activity transposes into more manageable terms traditional problems concerning the place of intentionality in nature" (Ibid: V, §64). For Sellars, *reasons* are occupants 'of nature' and not separated by any kind of ontological partition. As suggested, however, he does take our understanding of ourselves in nature as having two distinctive vocabularies, neither finding grounds for their epistemic status outside the conceptual—and neither finding conceptual solidarity with the vocabulary belonging to the conceptual other.

Along these lines, Sellars (1963) makes a distinction between the 'manifest' and 'scientific' images—two frameworks by which we come to understand ourselves and our relationship to nature. Roughly, the manifest image takes shape in our pre-history; it forms our ideals about what kinds of creatures we think we ought to be. It relies on a distribution of normative concepts and can describe our conditions in terms familiar to our duties and obligations to each other. The scientific image relies on our descriptions of the kinds of creatures *we are*, or at least the kinds of creatures we *think we are*. It exploits the results of our scientific theorizing and experimentation, relying on the language of science and of folk psychology. Joseph Rouse (2015) points out that the images are in irreconcilable conflict, but eventually the scientific image will displace the manifest (Rouse 2015: 8). Rouse's thought, consistent with Sellars's own, yields the idea that "[a]n adequate fusion of the images should give priority to the scientific image, situating our self-conception as sentient, sapient, rational agents within the horizons of a scientific conception of ourselves as natural beings" (Ibid: 8). McDowell's liberal naturalism depends on taking the order of nature and the order of reason to be part of the same world, but he seems to resist a fusion of images. His defence of the *sui generis* character of rationality relies on taking human beings as being fully natural creatures operating within the scope of the scientific image, but also requiring normative rationality to explain that which the scientific image cannot—and can never.

In "Liberal Naturalism and the Scientific Image of the World", David Macarthur (2018) suggests that "[l]iberal naturalism is a form of naturalism that positions itself to acknowledge the possibility of non-scientific entities or phenomena and non-scientific forms of knowing and understanding—categories that the scientific naturalist prejudicially and dogmatically regards as supernatural" (Macarthur 2018: 10). Macarthur takes McDowell to be a liberal naturalist in this sense, attempting to domesticate for our understanding aspects of human understanding that cannot be grasped by the conceptual powers of natural or empirical science. For Sellars and, indeed, for McDowell *all* awareness is by its nature a conceptual—or linguistic—affair (Sellars 1997: §29).[6] What one is aware of, and what concepts are exploited for one's is awareness, depends on what one is attempting to grasp. Where McDowell and Sellars share common ground is in Sellars's claim that "in characterizing an episode or a state as that of *knowing*, we are not giving an empirical description of that episode or state; we are placing it in the logical space of reasons, of justifying and being able to justify what one says" (Ibid: 76). Sellars's holism, wholly constitutive of rational norms, is also McDowell's—it is the locus of conscious awareness—*and of human freedom*.

McDowell's naturalized Platonism attempts to protect the *sui generis* space of reasons while bravely attempting to operate close to Quine's worldly tribunal. Echoing Quine's (1951) familiar pragmatic holism, McDowell writes: "experience must constitute a tribunal, mediating the way our thinking is answerable to how things are, as it must be if we are to make sense of it as thinking at all" (McDowell 1994: xii). It seems that the underlying naturalistic metaphysics of Quine's ontology is unproblematic (at least in its general form) for McDowell. However, his concern is with getting our thoughts to make contact with the world and then to have those thoughts line up with its constitutive features without giving any quarter to the bald naturalist or falling into the myth of the given. In order to allay this concern, McDowell gives an account of human nature.

McDowell believes that freedom, of the kind we associate with rational self-awareness, forms by way of an actualization of a *second nature* through education (McDowell uses the German term, *Bildung*). By education he does not mean *schooling*. Rather, he means a natural developmental process leading to the actualization of rational personhood. Rational norms relevant to such actualization are said to develop alongside the immediate "actualization of [our] conceptual capacities in sensory consciousness" (McDowell 2009: 470). So our exercise of freedom develops coextensively with the development of our sensory consciousness. The force of this claim finds its original articulation in Aristotle's discussion of phronēsis in *Nicomachean Ethics*.

In "Some Issues in Aristotle's Moral Psychology", McDowell (1998) thinks of orectic and doxastic states "as interlocking elements in a mechanism" (McDowell 1998: 30). Such an interlocking brings together both our sensory states and *inferential* states, something that is consistent with McDowell's own claim that there is not even a notional difference between receptivity and spontaneity (McDowell 1994: 9, 41, 51). Of course, he finds this insight in Kant as well (Ibid: 91). McDowell is attracted to Aristotle's account of practical wisdom such that the agent, in a particular situation, understands her actions in terms of intended ends and not merely as cases of an awareness of particulars. "So the premise of the good, and the selection of the right feature of the situation", writes McDowell, "serve as premise of the possible, correspond to a single fact about the agent, which we can view indifferently as an orectic state or as a cognitive capacity" (McDowell 1998: 30). This means that in any given situation an agent will act according to the ends she or he seeks by way of practical considerations—her desire and her beliefs presenting no explanatory advantage. McDowell identifies with the agent's wisdom a "true conception of the end" (Ibid: 30). So the ends the agent wishes to secure are *true* insofar as she or he understands those ends as shaping the outcome of the action in a way appropriate to the situation. Here McDowell relies on Aristotle's remarks in *Nicomachean Ethics* 1142, where Aristotle says,

> It is obvious that practical wisdom is not scientific knowledge; for, as we have said, it is concerned with the last thing, since this is what is done. It is therefore opposed to the intellect, since intellect is concerned with the first terms, of which there is no rational account to be given, while practical

> wisdom is concerned with the last thing; and this is the object of perception, not of scientific knowledge. (Aristotle 2009: 1142a23)

The kind of perception Aristotle seems to describe and upon which McDowell wishes to piggyback depends on rationalized ends that our acquired wisdom can inform of. By emphasizing the 'last thing' Aristotle is thinking of the practical side of wisdom—that is conceptually articulated and shaped by the ends the reasoning agent wishes to realize. Of course, language acquisition is key for McDowell because awareness of this kind is linguistic.

David Bakhurst (2015) in his 'Training, Transformation and Education' observes that McDowell premises the acquiring of a second nature, which McDowell equates with "a repository of tradition, a store of historically accumulated wisdom about what is a reason for what" on a generalized Aristotelian account of the formation of moral character (Bakhurst 2015: 305; McDowell 1994: 126). In addition, according to Bakhurst, McDowell exploits 'Gadamerian' themes that treat language acquisition as the transformational element in the development of second nature (Bakhurst 2015: 305).[7] Similarly, Courtney Fugate (2014) thinks that a neo-Aristotelian account of our *first* and second natures, underwrites McDowell's account of our initiation into language use. She argues that McDowell extends an idea of natural process into a *teleologically* fixed account of human actualizing of an ethical character (Fugate 2014: 62). McDowell has indicated as much in *Mind and World* arguing that as part of our natural history we acquire the "scrutiny of an ethical outlook" (McDowell 1994: 18). We might well imagine that the developmental process McDowell has in mind treats character formation and language acquisition as the mechanisms for a *transformational* initiation into the binding dynamisms of normative past—including inherited justifications for believing this over that and acting for this end over that.

Yet, McDowell is clear that an account of history or culture is not relevant to the developmental case he is making. His maturational strategy depends on the actualization of human potentialities already in the human organism that are fixed by their particular teleological ends. And those ends include a *truth-directedness with world*. This purely naturalistic *teleology* is not unlike the kind we might apply to particular species of flowers or African elephants or any other organism in a state of becoming. As such, we assess each by the ends it might successfully achieve or fail to achieve in its natural *unwilled* striving to be all that *it* can be as the kind of thing *it* is. McDowell thinks that having a second nature is *natural* in this sense. He writes that, "[h]uman infants are mere animals, distinctive only in their potential, and nothing occult happens to a human being in ordinary upbringing" (Ibid: 123). By the positing of *Bildung* he thinks he has provided a satisfactory account of how we are both natural and free, by *our* nature:

> Reflection about the *Bildung* of individual human beings should be enough to distinguish the naturalized Platonism I recommended from rampant Platonism. And in this reflection we can regard the culture a human being is initiated into as a going concern; there is no particular reason why we should need to uncover or speculate about its history, let alone the origins of culture as such ... And in any case, if we do speculate about how animals might have evolved into a way of living that includes imitating their young into a culture, we must be clear that that is what we are doing. It would be one thing to give an evolutionary account of the fact that normal human maturation includes the acquisition of a second nature, which involves responsiveness to meaning; it would be quite another thing to give a constitutive account of what responsiveness to meaning is. (McDowell 1994: 123–24)

McDowell's evolutionary story, including the idea that we 'imitate our young into a culture' amounts to a psychological account of upbringing that steers clear of historical or cultural considerations. McDowell is not concerned with *responses* to meaning, but with the descriptive claim that meanings require our responses—he is describing our capacity that by way of a general maturational process leads us *into rational personhood*. As such, McDowell's full person has a past without history. We can safely assume that here he is farthest from Collingwood.

Talk of a repository of tradition can give the impression that persons by McDowell's lights have histories. Instead, the repository of tradition, much like one might treat data that has been passed down from one generation to another, is part of the everyday and ordinary way by which norms can be binding. They are binding 'now' and justify our permissions 'today'. McDowell's maturational account treats the repository of tradition as a 'natural history' brought about by an evolutionary story about human development. Acquiring the *sui generis* character of belief holding and human action is simply an extension of a natural process. Given this, our maturational *telos* is distinctive insofar as humans are distinct from flowers and African elephants. This is not far from what a biologist might maintain about the development of the aortic valve in human mammals or a psychologist about dissociative disorders.

For McDowell, human beings are distinctive only in their potential—the majority will have healthy aortic valves and fewer than three percent (in Canada) will be diagnosed with a dissociative disorder, when the *Diagnostic and Statistical Manual of Mental Disorders-IV's* criteria are applied. And this sort of criteria, rather than residing 'in' reason itself, purports to help us adjudicate our judgements about the *facts* that are independent of reason—the sorts of facts that belong to nature itself. Still Sellarsian, even such facts by McDowell's lights are articulated conceptually and located in a logical space appropriate to their etiology. And such descriptions of our natures and the world are veridical. Just as African elephants *recognize* the appropriate plants that they require for nourishment, we acquire an orientation that corresponds to what is true. McDowell writes that:

> [w]hen one thinks truly, what one thinks *is* the case. So since the world is everything that is the case ..., there is no gap between thought, as such, and the world .... But to say there is no gap between thought, as such, as the world is just to dress up a truism in high-flown language. All the point comes to is that one can think, for instance, that *spring has begun*, and that the very same thing, that spring has begun, can be the case. That is truistic, and it cannot embody something metaphysically contentious, like slighting the independence of reality. (McDowell 1994: 27)

In his essay, "McDowell and identity theories of truth", Julian Dodd (1995) argues that McDowell's concerns with truth trade on a conflation of two distinctive and incompatible theories. While I do not intend to rehearse details of Dodd's argument here, nor am I interested in litigating Dodd's arguments against McDowell, I do think that it is important to consider McDowell's claim that to say there is no ontological gap between thought and world is to state a truism. McDowell's account of truth depends fundamentally on his account of our maturing into a second nature. But this claim itself depends on accepting a naturalistic outlook that treats the faculties of receptivity and spontaneity as naturally inclined towards the alignment of thought with world *because* the telos of thoughts is to align with world. When McDowell states that our conceptual capacities develop alongside our sensory consciousness, this is what he means.

In this respect, the world exerts rational constraints on our thinking by way of our senses.

Dodd correctly points out that McDowell's 'Tractarian' assumption that the 'world is everything that is the case' is a substantial philosophical claim. Dodd wonders why McDowell simply assumes that it is not controversial. Dodd further distinguishes between *robust* and *modest* identity theories of truth upon which he thinks McDowell's mind-world relation depends. He claims that robust theories of truth are favoured by correspondence theorists: "items with particular objects and properties are constituents whose totality makes up the world" (Dodd 1995: 161). If one identifies true propositions with such items, Dodd argues, any gap between representation and reality is eradicated (Ibid 1995: 161). He argues that G. E. Moore is this sort of theorist. On a common sense view, mind is in the business of grasping mind-independent entities, but more importantly, such entities are the datum of experience and not 'conceptual' in the way McDowell requires. Since McDowell is committed to defeating the myth of the given, this option is prima facie unavailable to him.

The thrust of Dodd's critique of McDowell is that the latter treats both the realism of common sense and the *senses* of Fregean Thoughts as truisms. And, if Dodd is correct, McDowell is conflating two different identity theories of truth. However, for my purposes here, I am not interested in whether McDowell can *defend* both these 'truisms', as he understands them (it is odd, of course, to have to defend a truism), but rather in his underlying interest in connecting a maturational account of human freedom which seems to accede to an idealism, on the one hand, and a realism of common sense, on the other.

These claims flow from McDowell's confidence in naturalism and that human nature is fixed. And his aim is clear: to emancipate us from the oscillation between bald naturalism and rampant Platonism. On one reading, rather than naturalizing idealism, McDowell has idealized naturalism insofar as his naturalistic account underwrites *sui generis* rationality.

5 Collingwood's Historical Human Nature

Collingwood may share McDowell's ontological monism, but his anti-naturalism favours *method* over ontology.[8] Collingwood holds that human nature is not specifiable once and for all. He combines this view with a procedural view of persons as *distinctively* historical beings.[9]

Collingwood famously argues in his *An Essay on Metaphysics*, explicitly in a chapter he calls 'Metaphysics Without Ontology', that there can be no science of pure being (EM: 17). Indeed, he claims that such a science is a contradiction in terms (Ibid: 11). While 'regular science' is about a specific subject matter to which certain methodologies apply, a science of metaphysics with no specific object and would not have any distinctive method. Collingwood thinks that the very idea of metaphysics as the study of pure being is nonsense; it is not even a pseudoscience or a quasi-science (Ibid: 15). In *An Essay on Metaphysics* he argues that the "problems of metaphysics are historical problems; its methods are historical methods" (Ibid: 62). By privileging *history* and *method*, and thereby favouring the centrality of normative rationality and human action, Collingwood is elevating history to transcendental status, "doing for the historical consciousness of to-day what Kant's transcendental analytic did for the scientific consciousness of the eighteenth century" (IH: 233). Here the 'historical' stands as a condition of knowledge, but not in relation to any particular *empirical* claim about this or that particular fact. Rather, the historical is a condition of conscious awareness itself—the kind of awareness that belongs to rational actors who are free. If one were to accept Kant's transcendental idealism, for instance, one would assume that geometrical facts are facts about space and arithmetical facts are facts about time. Neither space nor time are features of reality; they are features of mind and how mind experiences reality. In this way, for Collingwood, history is a condition of our experience and not a feature of 'reality'. History is also *our nature*. Collingwood offers a developmental starting point that is entirely different from McDowell's. Whereas McDowell's infant is to mature into the normative demands of reason and the civilizational inheritance locked in a "repository of tradition", Collingwood's infant is by her nature historical, developing into a person from being a person.[10] Put differently, Collingwood offers a twist to Hegel's dictum that *nature has no history* in asserting that "[m]an has, or rather, is history" (PH 1999: 199).[11] In *Principles of History*, in a section called 'Reality as History', he writes:

> What we call human nature is the product of historical processes, and the past as living in the present. Though what history has produced is mere fact and unalterable, 'nature' is something more than fact, it is compulsion: 'If I do actions of a certain kind because, as a matter of historical fact, I have

> acquired the habit of doing them, it is an unalterable fact that I acquired that habit: but it does not follow that, in further course of my history, the habit cannot be modified or broken. On the contrary, since the habit is mere fact, it falls away, like any other fact, into the past, unless it is constantly renewed by fresh action'. Thus if the habitual 'sets' or recurrent patterns of action in a given man or people are historically produced, the gates of the future are open; if natural they are shut'. (Ibid: 193–94)

We can distinguish clearly McDowell's view of human nature from Collingwood's. While historical processes act on us, they produce 'compulsions', the 'past lives in the present' through our acquired habits. Of course, as Collingwood notes, that acquisition of habits does not render them unalterable. Habits do not become part of a natural or evolutionary process. 'Fresh actions', justified and free, renew the actions. To think of them as 'natural' is to void their future. For Collingwood, then, human nature on this view is neither fixed nor does it have a fixed teleology. This said, given the status of history as a condition of our awareness, even nature is unfixed in that it belongs, in a substantive way, to our unfixed natures.

We describe nature by sequences of events and we generalize over what patterns we discern. When Collingwood writes, "[a]ll history is the history of thought" he is not excluding our understanding of nature (IH: 215). And the kinds of thoughts we have expressed and recorded, while in themselves particular to an instance, are of a piece with whatever is the state of mind of one who is recovering these thoughts and "re-thinking them in his own mind" (Ibid: 215). Still, the evidence of perceptive experience, observable and thought to conform to laws, is different in kind from historical 'events' that require for understanding the mediation of our rational faculties. So metaphysics is both a product of human freedom and *historical* in character; it is, in a manner, an inherently human type of intervention into understanding in its most generalizable form. Given Collingwood's dictum that *all* history is a history of thought, metaphysical questions are historical questions.

In *An Essay on Metaphysics* Collingwood stipulates:

> All metaphysical questions are historical questions, and all metaphysical propositions are historical propositions. Every metaphysical question either is simply the question what absolute presuppositions were made on certain occasion, or is capable of being resolved into a number of such questions

> together with a further question or further questions arising out of these. (EM: 49)

Collingwood argues in *An Essay on Metaphysics* that absolute presuppositions, as conditions of our experience, are not verifiable—nor can they be denied—they are neither true nor false (Ibid: 32). Since presuppositions are not *propositional* they are not truth-bearing. They have logical-epistemic status, much like geometrical facts in relation to space. So by saying that metaphysical questions *are* historical questions is to say that metaphysics is, by Collingwood's lights, a science of absolute presuppositions.

It is in In *Idea of Nature* that Collingwood claims that nature depends on history for its existence (IN: 176). Here he is not suggesting that nature is 'causally' dependent on history, but that nature is understood historically in that natural science is never presuppositionless. He illustrates his point: *natural science* comprises of 'facts' and 'theories' (Collingwood assumes a positivist science); a scientific fact is an event in the natural world; a scientific theory is a hypothesis about that event—and presumably future events will verify or falsify the theoretical efficacy of the prior event (Ibid: 176). Collingwood notes that a scientist in describing her or his observations of an event will exploit the vocabulary of science. Assuming that the observer of record has left a reliable testimony, a future interpreter will consult the record. This method is not unlike that of the historian who has to rely on her or his assumptions about the testimony of those who have left a record. That record demands interpretation; it is a record of events and not, of course, the events themselves. Where McDowell's realism privileges the ontology of nature as something we by way of our maturation discover, Collingwood domesticates science, as method, for his historical metaphysics.

Collingwood claims that "a 'scientific fact' is a class of historical facts; and no one can understand what a scientific fact is unless he understands enough about the theory of history to understand what an historical fact is" (Ibid: 177). Scientific theories are similarly dependent on certain historical facts—and they are "verified or disproved by certain other historical facts" (Ibid: 177). To be clear, scientific facts are historical facts. Metaphysics is a historical science because it studies presuppositions. Collingwood is treating the terms of a theory, its propositions, as parts of a historical record. Hence Collingwood's strongest claim is that 'natural science' as a "form of thought exists and always has existed in a context of history, and depends on historical thought for its existence" (Ibid: 177). For Collingwood, then, the language of science is historical—it depends on presuppositions. It is qualified and judgements belonging to rational willing shape its understanding of what it purports to describe of nature. Hence there cannot be presuppositionless knowledge since any appreciable judgements and discriminations that go into choosing this description over that belong to what can be conceptualized. Collingwood makes this point explicit in *Speculum Mentis* where he claims, "[i]n perception we are immediately aware of our object, which is a concrete and therefore historical fact: perception and history are just identical" (SM 2007: 205). He soon adds that "[p]erception is explicitly immediate (as perception is always contains within itself mediation (thought, 'interpretation of sense-data', 'inference from the immediately given', or whatever one likes to call it) and is therefore never abstract immediacy" (Ibid 205). Of note, here is Collingwood's claim that 'perception and history are just identical'. He means that perceptive experience is conceptualized and by way of its immediacy is inseparable from rational norms—or *history*. It is necessary, therefore, that Collingwood rejects non-conceptual contents as foundations for any claim, empirical or otherwise. So when Collingwood rejects givenness, his starting point is not an ontological 'gap' between mind and world but rather a positive claim about the nature of metaphysics as a historical science and history as a condition of our understanding.

6 Concluding Remarks

In *An Autobiography*, Collingwood rejects the 'intuitionist' theory of knowledge he ascribes to realists. Intuitionism, a theory often associated with G. E. Moore, favours *natural* intuition over reason and so is realist insofar as *truths* one intuits do not rely on inference or on normative judgements for their cognitive status—they simply are accepted as given. So we know that the intuitionist has no room for history—or rational norms. Collingwood thought intuitionism fundamentally flawed, not only because it seems to commit the naturalistic fallacy, but also that the intuited claim of the intuitionist stands outside of rational norms and yet is treated as inferentially foundational. The intuitionist assumes that the world independent of our will stands in relation to our thinking, but is, ultimately, left untouched by it. Given what Collingwood has said about nature, we can safely assume that he rejects realism. Perhaps more importantly, Collingwood's claim that 'all metaphysical questions are historical' is a challenge to what we take as 'given' at any particular time—including our ideas about our own natures. If Collingwood is right to assume that absolute presuppositions 'were made on certain occasions' he is also right to understand our sciences and, indeed, even our fashion choices as rational labours taking place in a context of a logical-epistemic given that one neither accepts nor rejects (A: 66). In other words, our inquiries and our assertions are subject to the historical conditions that support their efficacy. For instance, questions once thought entirely legitimate, whether women should have the right to vote or attend university in Canada, were informed by certain presuppositions that no longer have standing in our discussions as legitimate ways of understanding our moral or political order. This is not an assessment of the psychological or sociological dispositions of past actors who would exclude or include certain individuals. Of course, that is a legitimate study in its own right, but this is not our concern. Indeed, at the time of my writing this chapter, arguments marshalled to exclude certain groups from entry into the United States hold sway. Arguments for exclusion are of our moment, while questions about slavery or whether Indigenous people in Canada are full citizens seem settled. Still, it is important to keep in mind that absolute presuppositions establish the logical and not psychological conditions for the kinds of endorsements that obtain in a given context. So arguments for exclusion should be debated, but Collingwood's worries are not about what people do, but what they ought to do under certain conditions. D'Oro argues that to "this extent, like Kant, [Collingwood] is concerned with the question of right (*quid iuris*) rather than with the *de facto* question concerning the origin of knowledge" (D'Oro 2003: 6). She adds, "Collingwood's descriptive metaphysics, therefore, is not a sociology of knowledge" (Ibid: 6). In *The Idea of History*, Collingwood writes "the historical past, unlike the natural past, is a living past, kept alive by the acting of historical thinking itself, the historical change of one way of thinking to another is not the death of the first, but its survival integrated in a new context involving the development and criticism of its own ideas" (IH: 225–26).

McDowell's *liberal* naturalism rests on certain ontological presuppositions that treat mind and world as distinctive parts of the natural order. For him, history is *natural history* and he favours an account of *sui generis* norms and second nature that are, ultimately, extensions of a natural order. Collingwood's philosophy of historiography seems to bypass McDowell's concerns by taking our understanding of nature as part of our historical nature. Rather than our human nature arising out of 'nature', our understanding of nature is our responsibility and our *human* nature develops alongside our self-understanding, informed by different contexts and different renewals and rejections of our habits of mind. Collingwood's account of history and human nature is liberating, but it is not to be equated with McDowell's liberal naturalism.

Notes

1. Brian Keenan in conversation and in an unpublished presentation that identifies both Collingwood's claims that human nature is not specifiable once and for all combined with the procedural conception of persons as distinctively historical beings as central to the latter's philosophy.
2. We need not look further than Carl Hempel's (1942) influential "The Function of General Laws in History" for a naturalized account of history.
3. Aristotle (1933) considered Thales the founder of natural philosophy (Aristotle 1933: §983b20). While he did place deductive logic in the realm of mathematical reasoning, Thales is understood as being committed to a form of philosophical naturalism. Not entirely consistent with fully reductive views we associate with materialist ontologies today, Thales' outlook demonstrates that ontological naturalism can be traced (at least) to the pre-Socratic philosophers. We might also look at Callicles's excellent speech in Plato's *Gorgias* (482e–484d). Callicles defines the just in terms of nature's favouring the strong over the weak, famously arguing that this is demonstrated in nature as strong animals are superior to the weak (Schofield 2009). Not so much a defence of rational norms, Callicles defends instrumental instincts for survival where the powerful succeed and the weak are subjugated. Philip Kitcher (1992) argues in "The Naturalists Return" that *naturalism* was born in the nineteenth century and that everything before is pre-history (Kitcher 1992: 53; Rea 2002: 49). Kitcher traces contemporary analytic philosophy to Frege who has his philosophical progeny, including Russell, Wittgenstein, and Carnap (Kitcher 1992: 53). Kitcher's locating of the development of *contemporary* philosophical naturalism is not without merit. It speaks, however, to a narrowing of the discipline's historical record and the scope of its enterprise. Kitcher's main claims are outside the scope of this chapter, but his sentiment captures an attitude that is not uncommon.
4. Collingwood's criticisms of psychology are well documented. David Boucher's (1989) footnote in Collingwood's *Essays in Political Philosophy* notes of Collingwood that he took over 50 sessions of psychoanalysis before believing himself qualified to comment on Freud (EPP: 81).

Boucher also acknowledges Collingwood's recognition of psychology's usefulness; Collingwood is primarily critical of the field's claim to be a science of mind (Ibid.).

5. Collingwood's discussion of the 'overlap of classes' in, *An Essay on Philosophical Method*, distinguishes between history and science (EPM: 26). There Collingwood claims that "history concerns itself with something individual, scientific thought with something universal" (Ibid: 26). He thinks that philosophy is closer to science since it concerns itself with universals ... "and empirical science considers man as such, not, like history, this man distinct from that" (Ibid: 26).
6. Sellars's psychological nominalism, as he defends it in "Empiricism and the Philosophy of Mind," shapes McDowell's own thinking. In §31 Sellars insists that children's awareness of particulars, universals, facts, etc., depends on the acquisition of language (Sellars et al. 1997: §31).
7. In his essay, 'What Myth?' McDowell (2007) argues that his position is related to Gadamer's distinction between 'being oriented towards the world' and 'merely inhabiting an environment' (McDowell 2007: 346). McDowell thinks that there is a consistency here between Heidegger and Gadamer; he quotes the latter, "man's relation to the world is absolutely and fundamentally verbal in nature" (Gadamer 2013: 475–76; McDowell 2007: 346).
8. D'Oro claims that Collingwood is an *anti-naturalist* in her *Collingwood and the Metaphysics of Experience* (D'Oro 2002: 135). I agree with D'Oro that Collingwood is committed to transcendentals. She argues that "his commitment to transcendentals ... is firmly grounded in his defense of metaphysics as the study of the fundamental principles that regulate domains of inquiry" (Ibid: 135). She premises her remarks on the following observations: 1. Collingwood is not an eliminativist nor is he a reductionist, 2. Collingwood is not Cartesian and is not an ontological dualist, and 3. Collingwood is not a property dualist or a non-reductive physicalist (Ibid: 135).
9. D'Oro et al. (2018) in 'Non-reductivism and the metaphilosophy of mind' provide an account of Collingwood's defence of the autonomy of the human sciences. They identify Collingwood's commitment to explanatory pluralism, tracing its inspiration to Kant (D'Oro et al. 2018: 15). They show how explanations that appeal to norms are different in *kind* and not just in *degree* from those depending on "empirical regularities" (Ibid: 9).
10. In his *Principles of Art*, Collingwood (1938) suggests in his discussion of a child's cry, and the appreciable judgments underlying her distinguishing between her cries, *the use of language* but not of speech (PA: 236). This suggests to Collingwood that the child is enforcing her will on the world, in the absence of a plan or outcome. Collingwood uses the term 'language'

to "signify any controlled and expressive bodily activity, no matter what part of the body is involved" (Ibid: 241). Suitability or "pre-eminence of one [expression] over the other" will depend on the normative utility of the expression's development in a given historical and civilizational context (Ibid: 243–244).

11. In §295 of his *Phenomenology of Spirit*, Hegel writes: "In the case of self-conscious man the specific forms of consciousness constitute an ordered line of development, a necessary spiritual history. Organic nature has no such history: it falls straight from pure universality into the brute singularity of existence" (Hegel 1977: §295/535). This remark figures in Chapter V. of *Phenomenology of Spirit* "The Certainty and Truth of Reason". Here Hegel distinguishes, in part, between reason and nature, providing, in a contemporary idiom, an epistemic distinction between conceptual and non-conceptual content for the *observer* of nature.

Bibliography

Alexander, J. 2012. *Experimental Philosophy.* Cambridge: Polity.

Aportone, A. 2011. Rationalisms of Intuition, Kant and McDowell on the Relationship Between Conceptual Capacities and Sensory Experience. *Fogli Di Filosofia 2*: 75–89.

Appiah, K.A. 2014. Experimental Philosophy. In *Experimental Ethics.* London: Palgrave Macmillan.

Aristotle. 1933. *The Metaphysics.* Cambridge MA: Harvard University Press (Loeb Classical Library).

———. 2009. *Nicomachean Ethics*, ed. R. Crisp. Cambridge: Cambridge University Press.

Audi, R. 2011. Can Normativity Be Naturalized? In *Ethical Naturalism*, ed. S. Nuccetelli and G. Seays, 169–193. Cambridge: Cambridge University Press.

Bakhurst, D. 2015. Training, Transformation and Education. *Royal Institute of Philosophy Supplement 76*: 301–327.

Barrett, R., and R. Gibson. 1993. *Perspectives on Quine.* Oxford: Wiley-Blackwell.

Bedau, M. 1991. Can Biological Teleology be Naturalized? *The Journal of Philosophy* 88 (11): 647–655.

Collingwood, R.G. 1922. Are History and Science Different Kinds of Knowledge? *Mind*, New Series, 31 (124): 443–51.

———. 1924. *Speculum Mentis.* Oxford: Clarendon Press.

———. 1933. *An Essay on Philosophical Method*, revised ed. Oxford: Oxford University Press, 2005.

———. 1938. *The Principles of Art.* Oxford: Clarendon Press.

———. 1945. *The Idea of Nature*, ed. T.M. Knox. Oxford: Oxford University Press.

———. 1989. *Essays in Political Philosophy*, ed. D. Boucher. Oxford: Oxford University Press.

———. 1993. *The Idea of History* [1946], ed. J.W. van der Dussen. Oxford: Oxford University Press.

———. 1998. *An Essay on Metaphysics* [1940], ed. R. Martin. Oxford: Oxford University Press.

———. 1999. *The Principles of History*, eds. W.H. Dray and W.J. van der Dussen. Oxford: Oxford University Press.

———. 2013 *An Autobiography* [1939], eds. D. Boucher and T. Smith. Oxford: Oxford University Press.

D'Oro, G. 2002. *Collingwood and the Metaphysics of Experience*. London: Routledge.

D'Oro, G. 2003. *Collingwood and the Metaphysics of Experience*. Abingdon: Taylor & Francis.

D'Oro, G., P. Giladi, and A. Papazoglou. 2018. Non-Reductivism and the Metaphilosophy of Mind, *Inquiry*, special issue: 1–27.

Davidson, D. 2001. Actions, Reasons, and Causes. [1963] In *Essays on Actions and Events*, revised ed., 3–20. Oxford: Oxford University Press.

De Caro, M., and D. Macarthur, eds. 2010. *Naturalism and Normativity*. New York: Columbia University Press.

Dodd, J. 1995. McDowell and Identity Theories of Truth. *Analysis* 55 (3): 160–165.

Donagan, A. 1962. *The Later Philosophy of RG Collingwood*. Oxford: Oxford University Press.

Evans, G. 1982. *The Varieties of Reference*. Oxford: Oxford University Press.

Fugate, C.D. 2014. *The Teleology of Reason*. Berlin/Boston: Walter de Gruyter.

Gadamer, H.-G. 2013. *Truth and Method*. London: Bloomsbury.

Hahn, L.E., and P.A. Schilpp, eds. 1986. *The Philosophy of W.V. Quine*. Chicago: Open Court Publishing.

Hegel, G.W.F. 1977. *Phenomenology of Spirit*. Trans. A.V. Miller, ed. J.N. Findlay, Oxford: Oxford University Press.

Hempel, C.G. 1942. The Function of General Laws in History. *The Journal of Philosophy* 39 (2): 35–48.

Kant, I. 1998. *The Critique of Pure Reason*, Ed. and trans. P. Guyer and A.W. Wood. Cambridge: Cambridge University Press.

Kitcher, P. 1992. The Naturalists Return. *The Philosophical Review* 101 (1): 53–114.

Knobe, J., and S. Nichols. 2013. *Experimental Philosophy*. Oxford: Oxford University Press.

Macarthur, D. 2018. Liberal Naturalism and the Scientific Image of the World. *Inquiry* 9: 1–21.

McDowell, J. 1994. *Mind and World*. Cambridge, MA: Harvard University Press.

———. 2007. What Myth? *Inquiry* 50 (4): 338–351.
McDowell, J.H. 1998. *Mind, Value, and Reality*. Cambridge, MA: Harvard University Press.
———. 2009. *Having the World in View*. Cambridge, MA: Harvard University Press.
Murphy, T. 2014. Experimental Philosophy. In *Oxford Studies in Experimental Philosophy*, 325–368. Oxford: Oxford University Press.
Peacocke, C. 1992. *A Study of Concepts*. Cambridge, MA: MIT Press (Bradford Books).
Pradhan, R.C. 2015. Mind and World: From Soft Naturalism to Anti-naturalism. *Journal of Indian Council of Philosophical Research* 33 (1): 1–22.
Quine, W.V.O. 1951. Two Dogmas of Empiricism. *The Philosophical Review* 60: 20–43.
———. 1957. The Scope and Language of Science. *The British Journal for the Philosophy of Science VIII* (29): 1–17.
———. 1969. *Ontological Relativity and Other Essays*. New York: Columbia University Press.
———. 1981. *Theories and Things*. Cambridge, MA: Harvard University Press.
———. 1990. *The Roots of Reference*. La Salle: Open Court Publishing.
———. 1993. Response to Lauener. In *Perspectives on Quine*. Oxford: Wiley-Blackwell.
———. 2013. *Word and Object*. Revised ed., Introduction by Patricia Churchland, Preface by D. Føllesdal. Cambridge, MA: MIT Press.
Rea, M.C. 2002. *World Without Design*. Oxford: Oxford University Press.
Rouse, J. 2015. *Articulating the World: Conceptual Understanding and the Scientific Image*. Chicago: University of Chicago Press.
Schofield, M., ed. 2009. *Plato: Gorgias, Menexenus, Protagoras*. Cambridge: Cambridge University Press.
Sellars, W. 1963. *Science, Perception and Reality*. New York: Humanities Press.
———. 1979. *Naturalism and Ontology*. Atascadero: Ridgeview Publishing Company.
———. 1997. *Empiricism and the Philosophy of Mind*. Introduction by Rorty, R. and supplement by Brandom, R. Cambridge, MA: Harvard University Press.
Sellars, W., R. Rorty, and R. Brandom. 1997. *Empiricism and the Philosophy of Mind*. Cambridge, MA: Harvard University Press.
Smith, N.H., ed. 2002. *Reading McDowell: On Mind and World*. London: Routledge.
Sorell, T. 1991. *Scientism: Philosophy and the Infatuation with Science*. London: Routledge.
Tang, R. 2009. Conceptualism and the New Myth of the Given. *Synthese* 175 (1): 101–122.
Van Orman Quine, W., and L.E. Hahn. 1986. *The Philosophy of W.V. Quine*. Chicago: Open Court Publishing.

CHAPTER 9

Collingwood and the Philosophy of History: The Metaphilosophical Dimension

Jonas Ahlskog

1 Introduction

The best known and most discussed part of Collingwood's *oeuvre* is undoubtedly his contribution to the philosophy of history. During the past half century, thousands of scholarly works both inside and outside professional philosophy have assessed Collingwood's key concepts concerning historical knowledge and experience. *The Idea of History* (1946) is one of the most influential books of the twentieth century. In addition to being the most studied of his works, Collingwood himself also considered the philosophy of history to be the focal point of his intellectual legacy. The importance Collingwood attached to the philosophy of history is unmistakable when he speaks about his unfinished 'masterpiece' on the subject, *The Principles of History*, as the work "which my whole life has been spent in preparing to write. If I can finish that, I shall have nothing to grumble at" (PH: lviii). Unfortunately, Collingwood did not live to finish *The Principles of History*—his plan for that book was to deal in detail

J. Ahlskog (✉)
Åbo Akademi University, Turku, Finland
e-mail: Jonahlsk@abo.fi

K. Dharamsi et al. (eds.), *Collingwood on Philosophical Methodology*, Philosophers in Depth,
https://doi.org/10.1007/978-3-030-02432-1_9

precisely with the topics of the present essay. This topic is the metaphilosophical import of Collingwood's philosophy of history considered not only as an elucidation of history as a distinctive form of knowledge and thought, but addressing also issues about the ways in which philosophy of history matters for our understanding of philosophy as a whole. Given that Collingwood did not completely spell out the metaphilosophical aspect of his philosophy of history, the following essay requires both rethinking and extending Collingwood's trajectory of thought.[1]

This essay examines Collingwood's philosophy of history and its relation to philosophical analysis.[2] First, it is shown that Collingwood offers a unique and important account of the philosophy of history as an independent branch of philosophical inquiry with its own aims, questions, and subject matter. For Collingwood, the philosophy of history is essentially a philosophical inquiry concerning the fundamental concepts and principles that govern history as a form of knowledge and thought. Subsequently, I will situate Collingwood in relation to contemporary alternatives and argue that Collingwood's account must have an essential role in any forthcoming philosophy of history. In the latter part of the essay, it is argued that Collingwood's ideas have a far-reaching metaphilosophical import to the extent that his philosophy of history is *not merely* about the principles of historical research. On the one hand, it is correct to describe Collingwood's philosophy of history as an elucidation of the a priori concepts that guide history as a first-order science. However, it is equally important to understand that Collingwood's philosophy of history is also an effort to clarify the relation between historical and philosophical thinking by delineating the basic historical nature of human experience and understanding. I will argue that only by appreciating the interconnectedness of these aspects will it become clear how Collingwood provides an account that shows both (i) how philosophy of history is a distinctive branch of philosophical inquiry and (ii) that the philosophy of history has metaphilosophical relevance beyond the sphere of historical research.

2 The Subject Matter of Collingwood's Philosophy of History

Collingwood's elusive concept of history is the needle's eye through which every proper understanding of his philosophy of history must pass. Contrary to colloquial usage of the term, Collingwood's philosophical deployment of 'history' does not designate a position in time—the past—but a distinctive way of understanding reality. To understand a phenomenon historically, for Collingwood, is to understand that phenomenon as belonging to the category of human agency. As Collingwood makes abundantly clear on numerous occasions, the subject matter of history is *Res Gestae*, 'things done', and human actions are only properly so called if they are embodiments of thought. The characteristic feature of history does not reveal itself in the contrast between past and present, but rather in the conceptual distinction between the categories of action (history) and nomological event (natural science). Making human reality intelligible in terms of action is the quintessential feature of history, and it is, for Collingwood, a contingent matter whether the actions in question happened two seconds ago or 2000 years ago (cf. IH: 219).

Given that, for Collingwood, history is not synonymous with the past, it follows that 'historical explanation' is a term of art in his philosophy. Ordinarily, to speak of historical explanation or understanding means something like locating X in the spirit of the times, or viewing the significance of X in relation to subsequent developments. In everyday language, to explain historically means situating actions and events in a temporal context. Such contextualizing is also central for Collingwood, but it is only the means and not the end of historical explanation. To explain an action historically is, essentially, to connect in our understanding the action with the reasons for its performance. It is only possible to establish that connection if we are able to trace, or *re-enact* as Collingwood famously argues, the practical argument involved in the action based on the epistemic and motivational premises of the agent. This practical argument is the thought that the action embodies, and it is a thought that serves both as an explanation and as an identification of the action in question. For example, the actions of 'murder' and 'manslaughter' are distinguished and identified by the thoughts embodied in the action, not by physical descriptions of the material event. Equally, the reasons for an action are often a sufficient explanation of the action: 'why did you cross the street? – I bought today's paper'.[3]

A much-debated issue is the extension of Collingwood's concept of action. On several occasions, Collingwood seems to restrict history to a specific kind of action: "*Res Gestae* are [. . .] actions done by reasonable agents in pursuit of ends determined by their reason" (PH: 46). Several interpreters have used such passages to claim that Collingwood's ideas are relevant only for the narrow domains of rational actions in political and intellectual history. However, the significant distinction to be made is not between different kinds of action, but between actions and something else, such as impulse and knee-jerk reactions.[4] Rationality is not, contrary to 'selfishness', a predicate that actions may or may not possess, but a regulative idea that runs through *all* actions to a higher or lower degree. For Collingwood, to call something rational does not presuppose a fixed standard of desirable behavior, but involves only minimal conditions of intelligibility. This condition of intelligibility is set by our own ability to make sense of human behavior *as action* by making connections with the reasons for its performance. Something that lacks every form of reasonableness, meaning that its performance does not in any way follow from the epistemic and motivational premises of the agent, is not something that we would call 'action' but rather try to explain in causal terms through psychology, neurology, or biology. On this reading, the concept of rationality is internal to the historical mode of understanding and thus comparable with the role of the concept of nomological cause in (Newtonian) natural science.

Collingwood's (in)famous claim that all history is the history of thought has also created much puzzlement. In order to understand his claim, one must keep in mind that 'thought' does not necessarily mean 'deliberation'. In a narrow sense, all history is the history of thought simply by virtue of the fact that all history is about action, and an 'action' is by definition an expression of thought. However, this stipulation would certainly not be sufficient for claiming that history in the normal sense, that is, as an investigation of human affairs in the past, must be the history of thought. Historical research is, after all, often about aspects of human reality that were not the objects of deliberate action. For instance, historians may study economic trends to explain phenomena that were entirely opaque to the agents themselves. This is true, but it does not follow that such investigations could do without access to the 'thoughts' of the agents in a wider sense of the term. If a historian wants to understand the differences that economic trends made to the lived experience of the historical agents, then this understanding will be dependent on an understanding of the thoughts of the agents themselves. This dependence follows from the fact that their thoughts, and the internal relations between them, are an integral part of the lived experience under examination (cf. IH: 317).

Naturally, Collingwood's idea of a philosophy of history relates directly to his distinctive understanding of the concept of history. In philosophy, our aim is to discover what exactly it is that we are doing when we are engaging with the thought of the first order, such as history, natural science, art, or religion, respectively. Philosophy is thought of the second order and this means that it reflects on the relation between subject and object in first-order forms of thought. The task of a philosophy of history is, therefore, to make explicit the a priori concepts, principles, and categories that are implicit in historical knowledge and understanding. Centrally, this task involves the explication of some of the notions just mentioned, such as 'action', 're-enactment', and 'thought', but also concepts with direct methodological relevance for history as a professional discipline, such as 'evidence' and 'inference'. The aim of this philosophy is to clarify the concepts that make history as a form of experience and knowledge possible. For Collingwood, the philosophy of history is a branch of descriptive metaphysics and comparable to his conceptual clarification of other modes of experience and knowledge, such as art, religion, and science.[5]

Considering that Collingwood equates history with human agency, it should be obvious that philosophy of history will not be relevant only for professional historians. Rather, making reality intelligible in terms of action is, as Collingwood puts it, a necessary and universal aspect of the human mind (cf. Van der Dussen 1981: 311). Thus, the fundamental concepts of history as a science are continuous with our ordinary ways of understanding human behavior as the performance of different kinds of actions. This means that a philosophical clarification of the basic concepts of the science of history will, at the same time, serve as a clarification of the concepts belonging to the domain of agency in human life, and vice versa. Given that 'action' is already part of human existence, the philosophy of history will have an irreducible subject matter as long as human beings continue to make sense of their lives in terms of the concepts of human agency, such as motive, reason, purpose, meaning and so forth. Philosophy of history is, therefore, not important only for those who want to understand the practice of professional history, but for anyone wanting to understand the different ways in which reality is intelligible to us.

3 The Fundamental Role of Collingwood's Philosophy of History

In contemporary philosophy of history, Collingwood is often assumed to be somewhat outdated. The main goal for Collingwood was to distinguish the principles of history from natural science, and this is no longer a central concern for contemporary philosophy of history. Collingwood's philosophy, however, did have an important role to play in the philosophical debates following the publication of C. G. Hempel's classic paper, "The Function of General Laws in History", in 1942. In this paper, Hempel forcefully challenged the idea that historical explanation, or historical thinking as such, was in any way conceptually or logically distinct from the law-based explanatory model of natural science. In these debates, Collingwood's arguments for the autonomy of history served as a firewall against the attempt to reduce historical understanding to nomological forms of explanation. These debates abated abruptly already in the 1970s, at which time, the philosophy of history took a distinctively new direction. This new direction was a 'linguistic turn' in which the historian's uses of language was placed at the center of attention.[6] The key concepts for this new philosophy of history were no longer explanation and understanding, but narrative and linguistic representation.[7]

In fact, Collingwood writes comparatively very little about narrative and linguistic representation in his works on the philosophy of history. This is not, I believe, an accidental feature, but a choice motivated by Collingwood's idea of a philosophical subject matter. If 'narrative' simply means a linguistic tool for making temporal connections between events, then it is certainly a tool that may be used regardless of whether the subject matter is history, planetary motions or the nesting behavior of migratory birds. In addition, the use of narrative is not a necessary element in historical research since historians may use other, non-diachronic, forms of explanation to get their points across.[8] Not all history is written in narrative form, so neither should the philosophy of history center on the concept of narrative. Hence, narrowly construed, Collingwood's 'neglect' of questions about narration is entirely justified since literary representation is primarily a contingent methodological question and not a (transcendental) philosophical one about what history universally and necessarily is (cf. IH: 349–52).

However, if 'narrative' is used in a very broad sense, as it often is, then it can indeed become what Collingwood calls a philosophical concept—meaning a concept belonging to the necessary and universal features of a particular mode of understanding. In this philosophical usage, 'narrative' does not denote merely a linguistic tool, but the general process of sense-making relevant for every kind of historical research. For instance, the concept of narrative would denote the processes by which we understand one action as an intelligible response to an earlier action, or the principles underpinning the fact that we can make sense of an event in light of earlier developments. However, if 'narrative' is used to denote this generic ability to understand coherence and continuation, then it should be clear that Collingwood is actually the 'narrativist' philosopher of history *par excellence*—questions about such basic forms of sense-making are indeed what his fundamental concepts, from re-enactment to the historical imagination, centrally address. Thus, if 'narrative' is construed broadly, then Collingwood is certainly not outdated but the philosopher everyone in contemporary narrativist philosophy of history should read.

Still, there is no denying that Collingwood's main ideas are in conflict with the general direction of post-Hempelian philosophy of history. In contemporary scholarship, it is the idea of *retrospective description* that serves as the distinguishing feature of history, not action.[9] The main supposition behind this idea is, roughly, that the specificity of historical understanding is rooted in the use of 'narrative sentences'—a classical example of such a sentence is: "The Thirty Years War began in 1618".[10] Centrally, narrative sentences reveal truths about the significance of an earlier event in light of later events. Thus, historical knowledge contains elements that were not (logically) available to the agents themselves since truths about the significance of an event continue to accrue after it has happened. A 'historical event' is something that exists only under a certain description, and the availability of descriptions alters with our position in time. This entails that, with the passage of time, historians can use narrative sentences to create new events and novel descriptions in order to explain the past reality that is unfolding itself. Hence, contrary to Collingwood's ideas, the concept of temporality is placed center stage in history. As Arthur Danto tellingly wrote: "the whole point of history is *not* to know about actions as witnesses might, but as historians do, in connection with later actions and as parts of temporal wholes" (Danto 1965: 183).

It is a common but erroneous assumption that narrative sentences are specific to history. However, every past phenomenon, natural or human made, is the possible object of a description that assigns significance to earlier events in light of later ones. A natural scientist may say: 'the retreat of the glaciers since 1850 was the beginning of global warming' or 'The mosquito bite of John Doe in 1923 was the start of a pandemic'. Hence, it is simply wrong to suppose that retrospectivity is peculiar to the subject matter of historians. However, a more interesting fact is that every retrospective description in history will have to presuppose the form of re-enactive understanding that Collingwood places at the core of historical knowledge. The reason is that there can be no *purely* retrospective description of historical events. This is due to the fact that retrospective descriptions are created by linking together two temporally separated events, and for a linking to be possible there must already be an understanding of the separate events in themselves.

Obviously, the retrospective significance of the *assassination* of Archduke Franz Ferdinand was not available to contemporary witnesses in 1914. However, assigning retrospective significance to this event is certainly dependent on first identifying Gavrilo Princip's action for what it was, and this requires re-enactive understanding. Similarly, the much-discussed process of colligation in historical research must equally presuppose re-enactive understanding. For example, the application of the colligatory concept of 'thaw' in Soviet history is dependent on the fact that we can already understand the significance of certain actions for the agents involved.[11] The idea of a 'thaw' is closely associated with the relaxation of state control, and one can meaningfully apply the concept only to the extent that one is able to identify such a process in the lives of the agents themselves—whether they were aware of it or not. In fact, the literal meaning of 'thaw' denotes a process of softening of what was once frozen stiff, and this very meaning makes it an appropriate metaphor for describing the softening of state control. However, understanding the actual events in question as 'relaxation' or 'softening' must certainly be dependent on the fact that we can identify that actions were at a later date being performed that were previously considered to be prohibited by the state and, equally, that acts of reprisal did not follow from state officials. There was a change in conduct, and such changes are discernable only through the identification of action by re-enactment.

Retrospectivity is not the whole point of history, as Danto believed, but a perspective that is parasitic on re-enactive understanding and thereby knowing "about actions as witnesses might". Without the latter, there would simply be no events to describe in retrospective and colligatory terms. Without re-enactive understanding, our retrospective descriptions can never be more than arbitrary projections on the blank screen of an unknown past. However, it is equally wrong to claim that there is such a thing as a *purely* horizontal, and thereby non-retrospective, understanding of action in history. This is also contrary to Collingwood's main doctrine: retrospectivity is, in fact, a central element in re-enactment itself. To engage in re-enactment is to re-think the thought that past actions embody from one's own position in time. As Collingwood often emphasizes, this is a critical process and historians can know the past only if they are "firmly rooted in the present" (IH: 60). Historical knowledge, for Collingwood, is knowledge of the past in the present, not telepathic time travel (cf. IH: 175). The integral element in this form of knowledge, re-enactment, is a process in which we re-think past thoughts and thereby give them a new quality as one element "within a whole of thought that goes beyond it" (cf. IH: 448).[12] Thus, the accusation that Collingwood neglects the retrospective element in historical thought stems from the misconstrual of a key concept in his philosophy of history.[13]

Collingwood's ideas must have a fundamental role for any philosophy of history that aims to do justice to human agency in our understanding of history. This role is often not appreciated from failure to comprehend how deep Collingwood's ideas about history actually go. Collingwood does not simply offer a standard for the explanation of action, but a delineation of the concepts and principles that make historical knowledge and experience possible. To illustrate, the reality of the human past is not available to the historian as ready-made actions to be explained by whatever standard she chooses. In that case, re-enactment would be qualitatively on a par with the application of biological or psychological models for the explanation of action. This is simply not the case. For it is only by re-enactive understanding that we are able to view human reality under the aspect of human agency at all. The meaning of an action is not separable from the reasons for its performance, and this entails that the reality of the human past is accessible to us *as action* only as far as our re-enactments are able to reach. On this score, Collingwood's ideas are directly relevant for some of the most hotly debated themes in contemporary philosophy of history, namely, questions about the accessibility of the past and the ontology of historical objects.[14]

4 The Philosophy of History and Philosophy

As previously outlined, Collingwood offers a descriptive metaphysics concerning the concepts, principles and subject matter that make history, in his special sense of the word, specific and possible. Collingwood is thereby one among several seminal twentieth-century philosophers to engage with the project of distinguishing the human from the natural sciences. In this respect, there is clearly a connection between Collingwood's philosophy of history and the works of continental thinkers such as Wilhelm Dilthey and Hans-Georg Gadamer. However, perhaps the closest parallel in this respect is to Georg Henrik von Wright's arguments for separating the human from the natural sciences. Like Collingwood, von Wright attended to the logical distinction between different models of explanation and identified the human sciences with the practical syllogism and the natural sciences with the covering law model (von Wright 1971). Still, Collingwood's philosophy of history is essentially also *more* than an effort to demarcate 'history' on the map of human knowledge and experience. For Collingwood, the philosophy of history is important not only for understanding history, but also for philosophy as a whole. The rest of this essay is devoted to understanding this issue.

If Collingwood's philosophy of history were merely one branch of his descriptive metaphysics, then there would be no reason for giving any special status to history in comparison with, say, art or natural science. Still, it is a fact that Collingwood does give a special status to history. This special status has two distinguishable but connected aspects. On the one hand, the special status of history is a product of the development of human thought. Collingwood claims, famously, that the chief business of twentieth-century philosophy is to reckon with twentieth-century history (A: 79). The sense of this claim hails from Collingwood's idea that, sometime in the past centuries, historical thinking has gone through a Copernican revolution comparable to the revolutionary changes in man's thought about nature in the times of Newton and Descartes. Of course, this is not to claim that human beings started to make sense of reality in terms of action only quite recently. Rather, Collingwood's claim is that it is not until modern times that human beings began to understand their entire social world as a product of human actions in the past, rather than the result of divine forces, fate or some other non-human factor. In addition, a Copernican revolution came about with the insight that we can understand our past through a process of critical examination of evidence in which subject and object are internally related.[15] The primary expression of this revolution was, for Collingwood, the spectacular advance of historical research methods in the nineteenth century. In this sense, the metaphysical branch of 'history' has a special status as a ubiquitous but philosophically underexplored mode of understanding.

On the other hand, somehow connected with the above development, history gets a special status since Collingwood argues for a transformation of the very concept of philosophy in light of the progress of historical thinking. This is what Collingwood described as the need for a *rapprochement* between history and philosophy, and, at other times, as the creation of a new philosophy conceived from a historical point of view. In the *Idea of History*, Collingwood aimed at a philosophy of history in the narrow sense, meaning a philosophical delineation of the concepts and principles internal to history as a form of knowledge and experience. In contrast, the goal he set for the future was "a general overhauling of all philosophical questions in the light of the results reached by the philosophy of history in the narrower sense" (IH: 6–7). This would produce a philosophy of history in the wide sense, which I take to be Collingwood's label for a philosophical delineation of history that makes clear its relation to, and transformation of, traditional philosophy.

One can quite easily state what Collingwood did *not* mean by a philosophy of history in the wide sense. The above quotes, coupled with others in Collingwood's typically pointed style of writing, such as "all philosophy is the philosophy of history" (IH: 425); "philosophy is liquidated into history" (PH: 238) or "[a]ll metaphysical questions are historical questions" (EM: 49), have supported interpretations to the effect that the later Collingwood historicized philosophy into oblivion. This amounts to the supposition that Collingwood's new 'philosophy' reduces to historical examinations of different ways of thinking peculiar to different times. However, this interpretation is wide off the mark given the fact that, for Collingwood, philosophy was always a critical conceptual investigation with the aim of making explicit what we in some sense already implicitly know.[16] What could it mean to 'liquidate' philosophy as conceptual analysis? Surely, anyone attempting any such thing is dragged right back into conceptual analysis, and thereby philosophy, by the obvious response: 'what do you mean by liquidate?' Collingwood attempted no such suffocation of thought, and it is telling that his last writings constitute a *philosophical* defence of the idea of civilization. Thus, the important question is not whether Collingwood was a radical historicist, for he was not, but it is about how he thought philosophy should change in light of the development of historical thinking.

Collingwood wanted to liquidate a particular conception of philosophy.[17] The target was the idea that philosophy is a search for timeless essences, and that logic is an external and universal standard applied as a corrective to cases of actual human thought. In a seldom-quoted passage, Collingwood writes that the philosopher who engages with human life must accept human life and that philosophers "can only find in [human life] the reason which is in it already" (IH: 495).[18] However, this is not to deny that human thought is analyzable in terms of logical relations, such as truth, falsity, negation, implication, and so forth. The idea is rather that the correct application of logical concepts to any actual case of human thought depends on discerning the sense of those thoughts in the historical context of their employment. The opposite of this conception, Collingwood labels 'Aristotelian logic' and it is a form of thinking which assumes that one can assess the validity of any actual case of reasoning by merely attending to its form (ibid.: 253–4).

Collingwood's self-proclaimed revolution in logic, which was to show that the sense of propositions depends on historically situated question and answer complexes, aimed at exposing the flaws of Aristotelian formalism. Collingwood argued for a Socratic conception of logic in which the validity of judgments, propositions, and thought are dependent on their function as instruments in human dialogue. The aim was not to do away with formal logical relations, which would be nonsensical, but to show that every assessment of formal validity is parasitic upon grasping the sense of the relevant propositions and thoughts in the lived experience of actual human beings. However, this does not simply mean that history provides the raw material for logical analysis. Rather, more profoundly, Collingwood argues that the authority and sense of logical compulsion itself is a function of the historically specific reactions and responses of human beings in dialogue.[19] Philosophy, therefore, requires historical understanding if it is ever to discern the actuality of logical relations in human life.[20]

Collingwood's revolution in logic does not signal the invention of a new logic, whatever that could mean, but the detection of what Bernard Williams would call an 'impurity' at the heart of philosophical analysis. As Williams writes, if philosophy is to have anything important to say, then it must "address a lot more than philosophy" (Williams 1995: 148). Philosophy, thus, should never become the analysis of pure logical form, but must remain 'impure' from the fact that every philosophical analysis depends on understanding forms of knowledge and experience external to any a priori self-definition of philosophy's proper aims and methods (cf. Moran 2016: 317).[21]

It is possible to unpack this notion of impurity further by considering how Collingwood views the relation between philosophy and the subject matter of history. Arguably, a central aim in any philosophical analysis will be to gain a reflective understanding of our own ideas and motivations. Philosophy is, at least since the days of Socrates, inextricably linked with the quest for a knowledge of self that makes explicit what we in some sense already implicitly know. It is beyond doubt that Collingwood shared this idea of philosophy. However, self-knowledge is not only an aim in philosophy, it is also, as Collingwood argued, the condition without which other forms of knowledge cannot be critically justified and securely based (IH: 205).[22] Yet, in the quest for self-knowledge, philosophy cannot fend for itself and Collingwood claims "history is the only way in which man can know himself" (PE: 180). This interdependence of philosophy and history in the search for self-knowledge relates directly to Collingwood's ideas about the subject matter of history. Collingwood writes:

> Nature consists of things distributed in space, whose intelligibility consists merely in the way in which they are distributed, or in the regular and determinate relations between them. History consists of the thoughts and actions of minds, which are not only intelligible but intelligent, intelligible to themselves, not merely to something other than themselves [. . .] because they contain in themselves both sides of the knowledge-relation, they are subject as well as object. (IH: 112)[23]

This explicates an idea about the subject matter of history that goes well beyond the concept of *Res Gestae* and the necessity of re-enactment for the understanding of individual action previously discussed. Crucially, Collingwood makes a distinction between the constitutive features of two different kinds of objects of understanding: (i) phenomena explicable from an outside perspective and (ii) phenomena that are already forms of intelligence and embody an understanding of themselves as part of their very identity. The latter phenomena include not only the narrow domains of individual human action but also collective social phenomena at large. Unlike planetary motions or the atoms of nuclear physics, human practices and institutions, such as the family, property, science, art, philosophy and so on, enter the world with a conception of themselves as part of their very constitution.

For example, the practices and institutions of 'trade', 'money' or 'family' are what they are in virtue of the concepts and forms of thoughts shared by the participants. The practice of 'trade' involves understanding the specific relations of buyer and seller, the use of money involves thoughts about the relation between price and value, and the sense of the notion of a 'family' is not separable from particular, and historically specific, ideas about relations of responsibility between parent and child. The shared nature of the relevant concepts and thoughts contained in human institutions and practices are themselves constituted by the agreement in responses and reactions in the interaction among the agents involved. Thus, human practices and institutions contain an internal understanding without which they would not be the kind of human phenomena that they are. This internal understanding is often not explicit to the participants themselves, and historians can go beyond it in reflective interpretations of the phenomena in question, but any study that completely abandons the participant's internal understanding would simultaneously abandon the phenomena and turn into a study of something else.[24] In Collingwood's language, this is to say that 'mind' is an irreducible element in every part of the subject matter of history, from actions to institutions and cultural practices.

The human condition necessarily involves the inheritance of historically constructed practices and institutions from previous generations. In other words, history is, as Collingwood claims, an integral part of human experience itself (IH: 158). Most importantly, the starting point for anyone within this human condition is not one in which the meaning of our inherited practices and institutions, including their essential relations of power, will be transparent to the individual subject. In Collingwoodian terms, we enter the human world without knowing the ways in which the past is already included in the present.[25] Consequently, history *is* self-knowledge for Collingwood from the fact that it is only through historical understanding that we will be able to disentangle the past from the present, and thereby gain a more transparent understanding of the human condition to which our self-understanding already belongs.

The possibility of self-knowledge stems from our ability, in the process of historical understanding, to make discoveries in the realm of meaning. By discovering the distinct meaning of concepts and thoughts belonging to the actions and practices of people in other times and cultures, we also gain an understanding of the limits and shapes of our own ways of thinking. Such discoveries in the realm of meaning are, in turn, for Collingwood, ultimately secured by the logical possibility of critically rethinking the same thoughts that the practices and actions of others embody.[26] Consequently, Collingwood's idea of history *as* self-knowledge is not based on an empirical hypothesis—'people who know more about their past tend to know more about themselves'—but should be read as a conceptual point about the historical nature of human thought. History is the only way to self-knowledge because our present understanding of ourselves is already a historical product.

We have now articulated at least two related senses in which the philosophical delineation of history matters for philosophy. First, it is only by doing philosophy of history that we can recognize that our own thinking, philosophical or empirical, is the product of historically given institutions and practices. What philosophy gains from historical thinking, in this sense, is not raw material for the application of timeless philosophical thought, but an understanding of the concepts and procedures beyond which we cannot seem to get (cf. Williams 2006: 195–6). Historical thinking can unveil the ideas that are absolutely presupposed as conditions for going on as we do and show why this is also subject to change. Second, a philosophical delineation of the subject matter of history will also be required for understanding *in what way* philosophical thought relates to its own historical context. If one assumes that this relation is causal, meaning that thought is merely a function of its context, then determinism and historicism follows. However, as Collingwood wanted to show with his extensive arguments for the possibility of re-enactment, human thought *can* reach beyond the confines of its own historical context. The philosophy of history, thus, enables us to understand the ways in which human thought is simultaneously both free and subject to the historical framework of our common human condition. In other words, history gives to philosophy an understanding of its own impurity.

5 Conclusion

Collingwood's philosophy of history has two closely related sides. On the one hand, Collingwood's philosophy of history constitutes a branch of his descriptive metaphysics. The main aim for Collingwood in this respect is to answer the question: how is historical knowledge possible? His answer is provided by a philosophical delineation of the fundamental a priori principles and concepts that govern history as a form of knowledge and experience. In the earlier part of this essay, I showed how Collingwood gives an account of the philosophy of history as an independent branch of philosophical inquiry with its own aims, questions, and subject matter. In order to appreciate his account, it is crucial to recognize that Collingwood's philosophy of history is essentially a philosophy about human agency, not a philosophy about 'the past' per se. This particular subject matter does not mean that Collingwood's account is outdated in relation to the main concerns of contemporary philosophy of history. On the contrary, as I argued, Collingwood's philosophy has direct relevance for current debates concerning the role of narrative, retrospectivity, and ontology in historical research. In addition, Collingwood's main concepts will be very useful for constructing a radical critique of many philosophical assumptions that underpin contemporary positions in the philosophy of history.

On the other hand, Collingwood's philosophy of history also constitutes a critique of traditional conceptions of philosophy. Collingwood's aim was not only to delineate 'history' as an independent branch of knowledge, but also to show how the development of historical thinking should transform the very idea of philosophy itself. This would constitute a philosophy of history in the wide sense. The aim of such a philosophy of history is to improve philosophy in light of the progress of historical thinking, not to reduce philosophy to a branch of the history of ideas. Rather, Collingwood wanted to deconstruct the idea that philosophy deals with timeless essences and show that there is historical 'impurity' at the very heart of philosophical thought. However, an impure philosophy is still philosophy and not history.

In the last part of the essay, I have wanted to show that there is a *rapprochement* between philosophy and history exemplified in Collingwood's concept of self-knowledge. The main question for Collingwood's philosophy of history in the narrow sense—how is historical knowledge possible?—is necessarily tied to an even more fundamental inquiry: how is it possible for the mind to know itself? The metaphilosophical role of Collingwood's delineation of history is most obvious in his answer to this question. History is the only way to gain self-knowledge since it is only by doing history that we may discover the historical nature of our own ways of thinking. Thus, it is only by historical thinking that philosophy can attain a reflective attitude toward its own concepts and motivations. However, as Collingwood makes clear, the *rapprochement* between history and philosophy is a bilateral affair: "of all forms of thought, history is that which stands nearest to philosophy and most shares its spirit" (IH: 424). As I read him, Collingwood meant that historical research is never a mere inquiry about external objects and events but must necessarily involve a process in which we make explicit what we in some sense already know. History properly understood includes reflection upon itself, and in this way is like philosophy.

Notes

1. For Collingwood's plans to write about the metaphilosophical dimension of the philosophy of history, see (PH: xix).
2. My general understanding of this topic is much indebted to the systematic accounts of Collingwood's philosophy by scholars such as W. H. Dray, Giuseppina D' Oro, Jan van der Dussen, Rex Martin, Heikki Saari, James Connelly and Louis Mink.

3. This is connected with Collingwood's controversial claim that when we know what happened we also know why it happened. This claim is not as problematic as it is often made out to be. Problems arise only if we, erroneously, assume that Collingwood is claiming that any statement of 'what' is also always a sufficient answer to every 'why' question we may possibly pose.
4. In this sense, Collingwood introduces a technical concept of action (cf. D'Oro 2017: 220).
5. Giuseppina D'Oro has argued for this interpretation in (D'Oro 2002).
6. During the past decades, there are clear signs of a change away from narrativism. For a discussion, see (Ahlskog 2018).
7. A testimony to the contemporary vitality of narrativist philosophy of history is (Partner and Foot 2013).
8. This has lately been emphasized in (Kuukkanen 2015).
9. Arguably, narrativists and postnarrativists alike share this paradigmatic idea. Cf. (Roth 2016).
10. Arthur Danto introduced this concept in (Danto 1965).
11. For an extensive discussion of colligation using the example of 'thaw', see (Kuukkanen 2015: 97–130).
12. See also (PH: 223): "History means not re-thinking what has been thought before, but thinking of yourself as re-thinking it".
13. For a discussion of such accusations, see (Dray 1995: 308–9).
14. Chris Lorenz has argued that these are the main themes in the theory of history since the 1990s (Lorenz 2011). For a discussion of Collingwood's relevance for such themes, see (Ahlskog 2017).
15. This internal relation, however, is not finally understood until the development of scientific history based on re-enactment.
16. The Socratic spirit of Collingwood's idea of philosophy is prevalent throughout his writings. For a particularly clear expression, see (IN: 59–60).
17. The enemy was what Collingwood called Realism (cf. A: 147–8).
18. There is no support for interpreting this as an argument for uncritical acceptance. As Collingwood writes, the acceptance should be in the spirit that Margaret Fuller expressed in saying "I accept the universe" (IH: 494–95).
19. I take this to follow from Collingwood's idea of the logical efficacy of a question being dependent on a question-answer complex and absolute presuppositions. See (A: 37–9 and EM: 26–33).
20. Peter Winch later developed this conception of logic and reasoning (Winch 1972).
21. The following elaboration is indebted to Moran's article.

22. This follows from the fact that knowledge is a two-way relation dependent on the clear understanding of both subject and object.
23. In this passage, Collingwood is alluding to and agreeing with Schiller.
24. The most well-known and penetrating discussion of these issues is to be found in Peter Winch's work which is, in this respect, a direct continuation of Collingwood's philosophy of history.
25. The continental equivalent of this is the contention that the *They* (das Man) permeates the human condition (cf. Guignon 2017).
26. A clear contrast is Michael Oakeshott's constructivist philosophy in which the possibility of such discoveries is not clear. For an in-depth discussion of Collingwood's idea of re-thinking the same thoughts, see (D'Oro 2000).

Bibliography

Ahlskog, Jonas. 2017. R. G. Collingwood and the Presence of the Past. *Journal of the Philosophy of History* 11 (3): 289–305.

———. 2018. The Idea of a Philosophy of History. *Rethinking History: The Journal of Theory and Practice* 22 (1): 86–104.

Collingwood, R.G. 1945. *The Idea of Nature*. Oxford: Oxford University Press.

———. 1993 [1946]. *The Idea of History*, ed. Jan van der Dussen. Oxford: Oxford University Press.

———. 1998 [1940]. *An Essay on Metaphysics*, ed. Rex Martin. Oxford: Oxford University Press.

———. 1999. *The Principles of History and Other Writings in Philosophy of History*, ed. W.H. Dray and W.J. van der Dussen. Oxford: Oxford University Press.

———. 2005. *The Philosophy of Enchantment: Studies in Folktale, Cultural Criticism, and Anthropology*, ed. David Boucher, Wendy James and Phillip Smallwood. Oxford: Oxford University Press.

———. 2013. *An Autobiography and Other Writings*, ed. David Boucher and Teresa Smith. Oxford: Oxford University Press.

D'Oro, Giuseppina. 2000. Collingwood on the Re-enactment and Identity of Thought. *Journal of the History of Philosophy* 38 (1): 87–101.

———. 2002. *Collingwood and the Metaphysics of Experience*. London: Routledge.

———. 2017. Collingwood's Idealist Metaontology: Between Therapy and Armchair Science. In *The Cambridge Companion to Philosophical Methodology*, ed. Giuseppina D'Oro and Søren Overgaard, 211–228. Cambridge: Cambridge University Press.

Danto, Arthur. 1965. *Analytical Philosophy of History*. Cambridge: Cambridge University Press.

Dray, W.H. 1995. *History as Re-enactment: R. G. Collingwood's Idea of History*. Oxford: Clarendon Press.

Guignon, Charles. 2017. Self-Knowledge in Hermeneutic Philosophy. In *Self-Knowledge: A History*, ed. Ursula Renz, 264–280. Oxford: Oxford University Press.

Kuukkanen, Jouni-Matti. 2015. *Postnarrativist Philosophy of Historiography*. Hampshire: Palgrave Macmillan.

Lorenz, Chris. 2011. History and Theory. In *The Oxford History of Historical Writing: Historical Writing Since 1945*, ed. Axel Schneider and Daniel Wolf, 13–36. Oxford: Oxford University Press.

Moran, Richard. 2016. Williams, History, and the Impurity of Philosophy. *European Journal of Philosophy* 24 (2): 315–330.

Partner, Nancy, and Sarah Foot, eds. 2013. *The Sage Handbook of Historical Theory*. London: Sage Publications.

Roth, Paul. 2016. The Philosophy of History. In *The Routledge Companion to the Philosophy of Social Science*, ed. Alex Rosenberg and Lee McIntyre, 397–407. New York: Routledge.

Van Der Dussen, W.J. 1981. *History as a Science: The Philosophy of R. G. Collingwood*. The Hague: Nijhoff.

Von Wright, Georg Henrik. 1971. *Explanation and Understanding*. London: Routledge & Kegan Paul.

Williams, Bernard. 1995. *Making Sense of Humanity*. Cambridge: Cambridge University Press.

———. 2006. Philosophy as a Humanistic Discipline. In *Philosophy as a Humanistic Discipline*, ed. A.W. Moore, 180–200. Princeton: Princeton University Press.

Winch, Peter. 1972. *Ethics and Action*. London: Routledge & Kegan Paul.

CHAPTER 10

The Later Collingwood on Method: Re-Enactment and Abduction

Chinatsu Kobayashi and Mathieu Marion

To Jan van der Dussen

"See the value of imagination," said Holmes. "[...] We imagined what might have happened, acted upon the supposition, and find ourselves justified".
Arthur Conan Doyle, Silver Blaze.
My work in archaeology [...] impressed upon me the importance of the 'questioning activity' in knowledge.
R. G. Collingwood, An Autobiography

The recent rise of experimental philosophy, the renewed faith in the virtue of formal arguments in philosophy, and concomitant criticisms of the traditional 'armchair' practice has given a renewed impetus to debates about 'philosophical methodology' and the hope to find *the* method that would set philosophy on a proper course to firm conclusions that emulate those of science—a dream certainly constitutive of analytic philosophy since its inception. What possible contribution to this debate could we hope to find in Collingwood? As a sort of preliminary to answering this question,

C. Kobayashi (✉) • M. Marion
Université du Québec à Montréal, Montreal, QC, Canada
e-mail: marion.mathieu@uqam.ca

K. Dharamsi et al. (eds.), *Collingwood on Philosophical Methodology*, Philosophers in Depth,
https://doi.org/10.1007/978-3-030-02432-1_10

we focus here on the later Collingwood's[1] contributions to methodology, namely the 'logic of question and answer' and the concomitant notion of 're-enactment', because we think he has been misread. A proper reading of Collingwood on these would thus provide new grounds for evaluating his potential contributions to our ideas about philosophical methodology.

In chapters vii and viii of *An Autobiography* and chapters vi and vii of *An Essay in Metaphysics*, the later Collingwood suggested a '*rapprochement*' between history and philosophy. Not philosophy broadly conceived but history of philosophy, as his *proviso* made clear:

> … one aspect of the *rapprochement*, namely my demand that when philosophers thought about history of their own subject they should recognize that what they were thinking about was history, and should think about it in ways which did not disgrace the contemporary standards of historical thinking. (A: 77)[2]

This involved his trying to characterize a *single method for (history of) philosophy and history*, and we take it that chapter v of *An Autobiography* on the 'logic of question and answer' and chapters iv and v of *An Essay in Metaphysics* on presuppositions are his most developed attempts at discussing this issue. We shall talk here of *inquiry*, as we see Collingwood as primarily attempting to make explicit the method he followed as an archaeologist and as a historian of Roman Britain. We also take it that 're-enactment' itself is not that method—a supposed alternative to the 'hypothetico-deductive method' of the natural sciences—but the result of having applied that method.[3] In other words, re-enactment of a thought or propositional content comes about as the result of having reconstructed a complex of questions and answers of which it is the last step. This will be the focus of our contribution to Collingwood on 'methodology' in philosophy, as we are going to argue for a radical claim, according to which the later Collingwood has hitherto been largely misread, and that a proper reading should come from a better understanding of what he had to say on inquiry. And we are going to claim that it stands closest to 'abduction', understood as 'ignorance-preserving', according to a well-known interpretation by Dov Gabbay and John Woods (Gabbay and Woods 2005: 43; Gabbay and Woods 2006: 192).[4] To our knowledge, the first person to suggest a link between re-enactment and Peirce's abduction was Jan van der Dussen (van der Dussen 2007).[5] We found the

Gabbay-Woods Schema extremely helpful in order to pursue this idea. We first review the origins of the current view in the Hempel-Dray debate, then review the Gabbay-Woods interpretation of abduction while explaining how it partially fits Collingwood's writings, thus shedding light on what he was really up to. Our claim is thus that we should move away from readings of Collingwood in terms of a debate foreign to his own concerns, but in the second part, we wish merely to open up a new perspective, not to put forward a fully worked out alternative reading.

1 Collingwood and the Hempel-Dray Debate

We believe that the later Collingwood has been misread for the following reasons. Collingwood died in January 1943, just about one year after the publication of C. G. Hempel's 'The Function of General Laws in History', in the January 1942 issue of *The Journal of Philosophy*. By that time, Collingwood had suffered multiple strokes, he had resigned his professorship and left Oxford, and his attention was geared towards seeing *The New Leviathan* through publication, sometime in 1942. There is no reason to believe that he was even aware of Hempel's paper, and he certainly never wrote anything concerning the issues raised in it. This means, however, that there is no reason to see his writings on inquiry as some sort of contribution *avant la lettre* to the debate generated by Hempel's paper. But it is in these terms that Collingwood has been largely read.

According to the 'hypothetico-deductive method' or 'covering law model' in its simplest (and narrow) form (see Hempel and Oppenheim 1948), an explanation provided as an answer to a *why-question* resulting from some observation, say of a particular phenomenon a, would subsume it under covering laws: L_1, L_2, ..., typically of the form:

$$\forall x\big(A(x) \rightarrow B(x)\big),$$

with help of initial conditions I_1, I_2, ..., entailing the antecedents of the conditionals. One would thus 'explain' the statement $B(a)$, where a is our particular phenomenon, from such laws with initial conditions such as $I(a)$, that logically entail $A(a)$. The important fact to keep in mind here is that explanation is conceived in a purely *deductive* matter.

The model and its applicability to history have been subjected to numerous serious objections that need not be reviewed here. It is perhaps worth noting at least a well-known one by Michael Scriven (Scriven 1959: 454–55), reprised by Saul Kripke in a recently published paper, written in 1960, when he was an undergraduate:

> If we ascribe Cortez's explorations to his cupidity, the explanation cannot be an instance of a more general law, 'Greedy individuals will explore distant continents to gain gold', since the latter generalisation is plainly false. And if we attempt to modify the generalization to indicate what impelled Cortez to take an action other gold-seekers might have shunned, we soon find that so many clauses must be added that our generalization is applicable only to Cortez himself. (Kripke 2017: 14)

It is interesting to note that Kripke mustered the argument in support of Collingwood's view (citing IH: 234), adding in a footnote that it is Hempel's view which is being opposed, "whether or not Collingwood had Hempel in mind, or even was aware of his paper" (Kripke 2017: 14n.8). This caveat shows how the domain was dominated at the time by Hempel's proposal, and how Collingwood was read as having positioned himself within the debate it sparked.

Hempel himself seemed not to have been aware at the time of Collingwood's writings about re-enactment, and in arguing for the applicability of this 'hypothetico-deductive method' to historical explanations, he used as a foil what he called the "method of empathetic understanding", which he apparently knew from his own German-speaking tradition (Hempel 1942: 44).[6] According to Hempel, the view amounted to the idea that the historian

> … imagines himself in the place of the persons involved in the events which he wants to explain; he tries to realize as completely as possible the circumstances under which they acted, and the motives which influenced their actions; and by this imaginary self-identification with his heroes, he arrives at an understanding and thus at an adequate explanation of the events with which he is concerned. (Hempel 1942: 44)

One can certainly read Collingwood retrospectively, as having a view on this issue. Accordingly, opinions about Collingwood ranged from reading him as close as possible to, say, Dilthey and crediting him with an alternative method, for example, when W. H. Walsh spoke of "intuitive insight" (Walsh 1967: 58) or Patrick Gardiner of a "sort of telepathic communication" (Gardiner 1952: 39) or even of "an additional power of knowing which allows [historians] to 'penetrate into' the minds of the subjects of their study and take, as it were, psychological X-ray photographs" (ibid.: 128), to reading him as closely as possible to Hempel himself and crediting him with a method which is identical to the hypothetico-deductive method.[7]

It is true that Collingwood comes close to the above quotation from Hempel, when he wrote that the historian trying, say, to explain why Caesar crossed the Rubicon, must envisage

> ... for himself the situation in which Caesar stood, and thinking for himself what Caesar thought about the situation and the possible ways of dealing with it. The history of thought, and therefore all history, is the re-enactment of past thought in the historian's own mind. (IH: 215)[8]

Reading this passage in terms of the above 'method of empathetic understanding' would amount, however, to ignoring too many aspects of his philosophy, including his perceptive critique of Dilthey, on what we called the 'problem of transposition',[9] as well as the recognition of the essentially critical character of re-enactment: one does not merely 'relive', one *ipso facto* assesses critically.[10] One may add that such 'methodological' readings of re-enactment also contradict Collingwood's critique of what he called the "discredited copy-theory of knowledge" (IH: 284).[11] For these and other reasons, these readings have been largely abandoned and, by far the most influential approach since has been that of William Dray, who suggested a more astute reading in terms of 'rational explanation', in his doctoral thesis *Laws and Explanation in History* (Dray 1957a).

Dray used Collingwood's ideas in devising his own 'rational explanation' model, introducing it while discussing cases where there is no reason to believe that the agent actually 'calculated' by 'strict deductive reasoning' or that they even recited to themselves the explanatory calculation 'in propositional form':

> ... it is tempting to say that in such cases there is *no* calculation to be *re*constructed by the historian. But such an admission need not affect the main point: for in so far as we say an action is purposive at all, there is a calculation that could be constructed for it [...] No doubt there are special dangers involved in such construction after the fact. But although we may have to examine very critically any particular example, the point is that when we do consider ourselves justified in accepting an explanation of an individual action, it will most often assume the general *form* of an agent's calculations.

> Since the calculation gives what we should normally call the agent's *reasons* for acting as he did, I shall refer hereafter to this broad class of explanations as 'rational'. (Dray 1957a: 123–124)

This much underscores the key idea that such 'rational explanations' need not be obtained by some special method such as that of 'empathetic understanding', the postulation of the latter is thus inessential to Collingwood's viewpoint. What is important is simply that one can construct such a calculation.

Now Dray argues further that, when Collingwood used 'projection' or 'identification' metaphors, while speaking about the need to put oneself in the agent's position to understand why he or she acted in such and such a way, the point of his metaphors is to be "more plausibly interpreted as a logical one" (Dray 1957a: 128; see also Dray 1995: 54). The idea is not just that the point of the metaphors is not to tell us something about how we come to know why he or she acted in such and such a way, but that, for an explanation to be successful it has to show that this was the appropriate action to perform and that this would be achieved via an appeal to the above kind of rational calculation. Given a distinction between reasons and causes, this brings his reading of Collingwood close to the nowadays much discredited 'logical connection argument'.[12] To us, this would be a 'no go' area, as well as an indication that Collingwood has been drafted in a debate which is foreign to his thought. The proper interpretation of Collingwood on this point would require more space than is available here, and we shall limit ourselves to a single argument, that, we assume, has hitherto not been made in the secondary literature.

Recall that Dray developed the above thoughts in 'rational explanation model' involving what he called 'principles of action', of the form: "When in a situation of type $C_1, \ldots, C_n$, the thing to do is x" (Dray 1957a: 132), so that we have the following type of explanation:

> A was in a situation of type C
> When in a situation of type C the thing to do is x
> ———————
> A did x.

One aspect that Dray emphasized is that "in an important sense, rational explanation falls short of, as well as goes beyond, subsuming a case under a general empirical law" (ibid.: 131), so that 'principles of action' are in that sense *generalizations* that are *weaker* than empirical *laws*. One is still talking here in *deductive* terms.

Of course, this move has been contested, starting with Hempel, who proposed to replace this sort of premise with premises of the form "Any person who is disposed to act rationally will, when in a situation of type *C* invariably (with high probability) do *x*", and revise the 'rational explanation model' accordingly, essentially bringing it in line with his own 'covering law model' (Hempel 1966: 117–18). To our mind, these developments show once more how far we have moved from Collingwood's thought and its context.

We argue this point from chapter 14 of *The New Leviathan*, where Collingwood examines three possible answers to the question 'Why did you do that?': 'because it is useful', 'because it is right' and 'because it is my duty'. This is the same as asking oneself, when confronted with a variety of possible actions in a given situation: 'Which shall I do?' His line of argument—our point here is not to evaluate it as such but simply to show that it reveals him to be at odds with the Hempel-Dray debate—consists in claiming that the various criteria, utility, right, and duty, act as *sieves*, so to speak, that are applied in order progressively to eliminate possible courses of action among a set of them given at the outset. One must thus imagine an agent having to decide what to do in a given situation, that she understands to be such and such, who then envisages as many choices as possible, only to use these criteria to eliminate less than optimal courses of action.

Collingwood then proceeds to argue that an element of caprice is left with utility or right, but not with duty, because in the first two cases some alternatives will remain open, and any choice between the remaining alternatives would then be, in the absence of any further criterion, purely capricious. On this basis, Collingwood was able to criticize utilitarianism and Kant's 'regularian' approach as unable to provide a proper account of duty. For example, in the latter case, rules only tell us that acts of a given kind are permissible, not which particular action should be undertaken, and, furthermore, there are often conflicting rules that apply to a given situation and that conflict cannot be resolved by a mere appeal to the rules. So, the 'regularian' account leaves open some alternatives. The peculiarity of duty is, according to Collingwood, that there are no alternatives left to choose from:

> Duty admits of no alternatives. Whatever is my duty is an *individuum omnimodo determinatum*. There is only one of it; it is not one of a set of alternatives; there is nothing that will do as well. In the first place it is my duty and nobody else's. [...] Secondly, any duty is a duty to do 'this' act and only 'this', not 'an act of this kind'. [...] Hence dutiful action, among these three kinds of rational action, is the only one that is completely rational in principle; the only one whose explanations really explain; the only one whose answer to the question; 'why did I do that action?' [...] answers precisely that question and not one more or less like it. (NL: 17.51–55)

This, Collingwood adds, does not preclude an analysis in terms of utility and right, but *includes* it (NL: 18.6 and 478). How does all this translate into historical thinking, that is, Gibbon on Caesar or Collingwood on Gibbon on Caesar? As it turns out, Collingwood believed that 'The consciousness of duty is thus identical with the historical consciousness' (NL: 477). Or, to cite another passage:

> The consciousness of duty means thinking as an individual or unique agent, in an individual or unique situation, doing the individual or unique action which I have to do because it is the only one I can. To think historically is to explore a world consisting of things other than myself, each of them an individual or unique agent, in an individual or unique situation, doing an individual or unique action which he has to do because, charactered and circumstanced as he is, he can do no other. (NL: 18.52)

This is not the place to assess the claim, which is doubtlessly strong: duty is such that, in any given occasion, for a given person there is only one way to act. Thinking instead in terms of the debate between Dray and Hempel on the nature of explanations in history, the upshot here is that Collingwood, had he lived to know about Hempel's covering law model, would have probably identified it with Kant's 'regularian' account, and would have claimed that, *alone*, covering laws cannot provide a proper explanation: something *more* is needed to give a full explanation, and not some *weaker* principle such as 'principles of action' proposed by Dray. Because his claim here is not that rational explanation will not include any kind of law or generalization: the sieve will ensure that any remaining possibilities include them. Now, the point is not that an argument such as Scriven's, above, would be welcomed, but that Collingwood would rather not be averse to using either covering laws or weaker generalizations. That covering laws would be involved or not is beside the point: he was after something else.

To try and shed light on this, we will suggest an alternative reading that does not ignore the bulk of his writings on the subject, in particular, chapter v of *An Autobiography* on the 'logic of question and answer' and chapters iv and v of *An Essay in Metaphysics* on presuppositions. Although we use here Conan Doyle's *Silver Blaze* (Conan Doyle 2008: 1–22) for the purposes of illustration, 'Historical Evidence', originally published as one of the 'epilegomena' to *The Idea of History* (IH: 249–82),[13] is also crucial here, especially given that it includes a very brief detective story *Who killed John Doe?* as an imaginary example on which one is to utilize one's intuitions to see what it reveals (IH: 266–68; PH: 21–24).[14] Part of the difficulty in reading Collingwood from the standpoint of the Hempel-Dray debate is that its participants were looking for an explanation model in which steps are *deductive*. There is ample evidence in the texts just mentioned that Collingwood did not think in such terms and that a model of *abductive* reasoning such as the Gabbay-Woods Schema is a more appropriate a tool for the study of his philosophy. After all, Collingwood was an archaeologist and a historian, and he admitted himself that his philosophical ideas, including that of 're-enactment', were derived from his practice as archaeologist and historian.[15]

2 The Logic of Abduction

The *locus classicus* for abductive reasoning is C. S. Peirce's seventh Harvard Lecture, 'Pragmatism as the Logic of Abduction' (1903), in which he argued that it is triggered by an element of surprise, and that it has to do with the formation of explanatory hypotheses. It would thus take the form:

> The surprising fact, *C*, is observed;
> But if *A* were true, *C* would be a matter of course.
> Hence, there is reason to suspect *A* is true. (Peirce 1998: 231)

The triggering element will be described below as an 'ignorance problem'[16]; it is a truism that problems of this sort typically form the starting point of archaeological and historical research, as well as they do for Sherlock Holmes or Hercules Poirot, to keep to the 'detective image of the historian' that many see at work in Collingwood.[17] This is our first cue.

In this section, we would like simply to review the Gabbay-Woods Schema for abductive reasoning, pointing out aspects of Collingwood's text that are enlightened by it. Our task here is merely to state the prima facie case, hoping that others will travel further down the path(s) thus opened. Of course, this Schema is only one of the many approaches to the logic of abduction[18] or to Peirce's text,[19] but it is an influential one, and we are interested in it inasmuch as it helps us understand what Collingwood was after, and we are not arguing for a perfect fit either.

To introduce the Gabbay-Woods Schema, we switch to a more modern and analytical jargon. It is useful to think in terms of an inquiry triggered, as we saw, from a question Q that arises in a given context, and for which a sentence α, if known, would be the answer or would help providing one. Assuming from the reader knowledge of Conan Doyle's short story *Silver Blaze*, let us take as an example of a triggering event that the eponymous horse has disappeared and that the stable master, John Straker, was found dead in the moors. 'Who killed Straker?' is one of the questions for which Sherlock Holmes' current knowledge base or its immediate successor do not provide an answer—this is why it is an 'ignorance problem' (Gabbay and Woods 2005: 42; Gabbay and Woods 2006: 191). There are three possible scenarios for Holmes with respect to Q:

(i) He obtains new knowledge that removes his ignorance, say, of α as the answer to Q (this is called 'subduance'),
(ii) He gives up looking for an answer,
(iii) He sets in the course of his investigation a hypothesis as the basis for further reasoning or knowledge-seeking actions.

This last is 'abduction', that is, an inference by which the ignorance problem is not solved as such, but *partly* overcome by conjecturing the hypothesis. As Woods puts it, it is "a response that offers the agent a reasoned basis for a new action in the presence of that ignorance" (Woods 2013: 368). This means that the hypothesis thus conjectured is neither necessarily true, nor a piece of knowledge: it is merely used in further reasoning or actions, and the initial ignorance is preserved in any inference from it, so that that abduction is best understood as an 'ignorance-preserving relation'.

More formally, the Schema goes as follows,[20] with these symbols:

- T stands for an epistemic agent's (such as Holmes) state at a specific time, the target.
- K is that agent's knowledge base at that time, and
- K^* an immediate successor of that knowledge base.

- R is an attainment relation for T so that $R(K, T)$ says that knowledge base K is sufficient to attain target T.

So let us now set T as a target which is to get to know α, which would be the answer to Q and write it thus:

$$T!Q(\alpha). \tag{1}$$

It is part of the definition of being an 'ignorance problem' that knowledge bases K and its immediate successor K^* are presumed to be insufficient to reach T. So, we have (with '¬' standing as usual for negation) both:

$$\neg R(K, T). \tag{2}$$

$$\neg R(K^*, T). \tag{3}$$

Indeed, if K^* were sufficient to attain T, Holmes would merely need to extend his knowledge, it would be 'subduance' as in (i) above, and there would be no need for abduction.

Let us now call H the hypothesis that Holmes would need to conjecture in order to formulate a solution to the ignorance problem. It cannot be, for the reason pointed out above, a true sentence or a piece of knowledge, and this is represented as follows:

$$H \notin K. \tag{4}$$

$$H \notin K^*. \tag{5}$$

Even combining it with the knowledge base K, as in $K(H)$, would not allow one to attain T, so:

$$\neg R(H, T). \tag{6}$$

$$\neg R(K(H), T). \tag{7}$$

Why bother with the hypothesis H then, if it does not settle the ignorance problem? It is here that Gabbay & Woods interpret Peirce's schema above, by saying that the hypothesis *subjunctively* relates the knowledge base K to the target: it means that H is not a true sentence or a piece of knowledge, but if it were, it would provide an acceptable solution to the ignorance problem. This is captured as follows, with help of a subjunctive conditional connective, here '$\Rightarrow$', for which no particular formal interpretation is given:

$$H \Rightarrow R\big(K(H),\, T\big). \tag{8}$$

This hypothesis is thus a 'subjunctive' solution.[21] Putting the premises (1)–(7) together one can conjecture the hypothesis, or rather *conclude* that we are right in conjecturing it:

$$C(H). \tag{9}$$

Recall here that abduction, as in (iii) above, is not equal to the conjecturing of a hypothesis: it is setting it as the basis for further reasoning or knowledge-seeking actions. So, a further step is needed, that of 'releasing', so to speak, the hypothesis for further reasoning or action in what would be abduction proper or 'full abduction'. With (9) we have only a 'partial' abduction.[22]

Before we get to full abduction, let us see what possibilities there are once the hypothesis is conjectured:

I. Supposing it is tested and it is confirmed, we have (a) new knowledge (as in K^*), we thus have what was called above subduance. So, we do not act on the hypothesis in an ignorance-preserving way—there is no full abduction—but we have possibly something like hypothetico-deductive reasoning. Supposing it is tested but not confirmed, we could (b) either give up the hypothesis, ending in partial abduction or (c) nevertheless keep the hypothesis and perform full abduction.
II. We could also *not* test the hypothesis and 'release' it for further reasoning, that is, go for full abduction. With the superscript indicating the conjectural origin of the hypothesis, this would amount to concluding from (1) to (8):

$$H^C. \tag{10}$$

Sherlock Holmes' stories are replete with examples of 'full abduction', and in *Silver Blaze,* the "curious incident of the dog in the night-time" illustrates it very well:

> "Is there any point to which you would wish to draw my attention?"
> "To the curious incident of the dog in the night-time"
> "The dog did nothing in the night-time"
> "That was the curious incident," remarked Sherlock Holmes. (Conan Doyle 2008: 17)[23]

When unpacked, the reasoning behind this episode is as follows: Silver Blaze was taken away from the stable at night-time, but the guard dog did not bark in presence of the thief. However, the generic 'guard dogs do not bark at their masters' is good enough to conjecture that the stable master, Straker, might have been the thief. So, *without knowing this*, Holmes pursues his investigation accordingly, seeking further evidence. It is furthermore all too easy to think of examples in terms of archaeology, given that full abduction appears to be enshrined in its very method of 'selective excavation'.

At the end of Sect. 1, we suggested that abduction would help us seeing why Collingwood's interests are at best tangential to the Hempel-Dray debate. The contrast we should focus on in this respect is between testing and confirming the hypothesis in I(a), which results in a piece of hypothetico-deductive reasoning, and the cases of 'full abduction' in I(c) and II, when one reasons from or acts on a hypothesis, which is recognized as neither a true sentence nor a piece of knowledge, in an 'ignorance-preserving way'. We think that this contrast allows us better to understand Collingwood, who is clearly interested in instances of the latter, at the expense of making his views as clear as possible on the former. For instance, one might simply recall his view of 'constructive history' in terms of "acts of interpolation" (IH: 242–45), such interpolations being instances of full abduction in the historian's work. It is worth noticing here, since we mentioned 'selective excavation', that Collingwood specifically linked the origin of his 'logic of question and answer' to his practice as an archaeologist:

> ... long practice in excavation had taught me that one condition – indeed the most important condition – of success was that the person responsible for any piece of digging, however small and however large, should know exactly why he was doing it. This was the central principle of my 'logic of question and answer' as applied to archaeology. (A: 122)

In his capacity as archaeologist, Collingwood's mentor was F. G. Simpson, who pioneered (with J. P. Gibson) the method of 'selective excavation'.[24] This may have resulted from practical necessity, as on a given site—an obvious case at hand being Hadrian's Wall—complete excavation is impossible, so one must conjecture hypotheses and then select particular spots for digging (thus 'releasing' the hypothesis for further action), hoping that what will be revealed not only will confirm the hypothesis, but possibly raise further questions for further selective excavations. Moreover, as in I(c) above, cases abound in archaeology where the hypothesis is maintained even if evidence from a selective excavation does not confirm it.

In relation to this, one must underline some of the things Collingwood wrote about the 'logic of question and answer', especially with regards to the idea that to see the point of a given proposition, one must recover the question to which it was an answer, and possibly reconstruct the whole 'complex' (A: 37) of questions and answers behind it, a sort of chain where the answer to a question forms a presupposition to another question, and so forth. One of the key points of the chapter is that any answer α to a question Q is not 'true', but 'right' inasmuch as it allows one "to get ahead with the process of questioning and answering" (ibid.). One should note that this is not exactly as above, where it is the hypothesis H, not α, which is recognized as neither being true, nor a piece of knowledge. The fit is thus not perfect, but it is worth noting that in the 'logic of question and answer', an answer α is the trigger that leads to conjecturing the question Q, which is then treated abductively as H^C above. This is why Collingwood insists on 'going ahead with the process' as opposed to truth.

3 Concluding Remarks

We would like simply to conclude with two supplementary remarks, first concerning chapters iv and v of *An Essay in Metaphysics* on presuppositions, and, second, providing further emphasis on the fact that Collingwood wasn't truly interested in 'method' in the sense at work in the Hempel-Dray debate.

So far as history of philosophy is concerned, Collingwood argued in *An Essay in Metaphysics* that one would need to reconstruct the complex of questions and answers up to absolute presuppositions. In the context of our chapter, it is worth noting that, although the result of one's investigations might possibly be neatly reordered in some 'axiomatic' fashion, with absolute presuppositions standing at the beginning of the chains of questions and answers, in practice the historian of philosophy is thinking in terms of full abduction, raising questions such as: 'Why did X write α?', and conjecturing the question Q to which α is the answer, in order to set out for further clues in the text. This is the process that, we would claim, Dagfinn Føllesdal misunderstood, when he tried to fit it in the mould of Hempel's 'hypothetico-deductive method' (Føllesdal 1979). It is also worth pointing out in this connexion, that Collingwood saw this process as open-ended: in history "no achievement is final" as new evidence might come to the fore and new questions are asked by each new generation, we only have interim reports (IH: 248), and in art (after all, Føllesdal's example concerns the interpretation of one aspect of Ibsen's *Peer Gynt*), because there is no definitive meaning of the work of art to be captured (PA: 311). Of course, both claims raise fresh issues, but the point here is merely to see that Collingwood had in mind an open-ended process, where ignorance is recognized and preserved all along.

Perhaps a comment on Collingwood's own detective story *Who Killed John Doe?* is in order, to emphasize the very last point of the previous section. This was the idea that one infers a question Q from a set of premises X, a question that would then be handled abductively. Now, after telling the very brief *Who Killed John Doe?* Collingwood drew some conclusions. The first one of them is that "every step in the argument depends on asking a question" (IH: 273). This statement can be read as being of the effect that one would not infer Q solely from a set of premises X, but also in conjunction with another initial question Q^*, so that we would have an (ordered) triplet $<Q^*, X, Q>$, instead of the (ordered) pair $<X, Q>$.[25] This is tantamount to thinking that one reasons from question to question, without any appeal to further premises.

We sometimes read about what is perhaps abusively called the 'phenomenology' of rule following, for example, with the example of the experienced chess player, who does not think in the midst of a game of the rule for moving the Queen in order to verify if he will be moving it accordingly, but just moves it somehow without thinking about the rule. This is of course a contentious point to make against those who think that awareness of the rule is always involved. We obviously do not wish to enter this debate, but we note that something of the sort is at play with Collingwood, which would be, in parallel, an instance of the 'phenomenology' of abductive reasoning in inquiry. This is a very large question, which would deserve further study, so that we would deepen our understanding of Collingwood. But for the moment raising the issue should suffice further to show that Collingwood's own interests were far from those of the participants to the Hempel-Dray debate.[26]

Notes

1. We would like provisionally to define Collingwood's later period, as beginning at the time of his appointment as Waynflete Professor of Metaphysics in 1935. We make no claim concerning his earlier views on philosophical method, as well as no attempt to draw links or to suggest a unified reading of his philosophy. Not that we do not believe one to be possible, but merely because we wish to keep the topic of our contribution well circumscribed.
2. One subsidiary question not to be tackled here is that of the extent to which Collingwood thought of 'metaphysics' as merely 'history of metaphysics'. This would be a reductive reading of Collingwood. See (D'Oro 2002) for contrary evidence.
3. This claim is rather common since Donagan's claim that "Collingwood's subject may be, not historical method, but what historical method achieves" (Donagan 1956: 203), and it does not seem in need of new arguments. See the survey in (van der Dussen 1981: 96–109) and, for example, (Ricœur 1984: 6) or (Dray 1995: 55) for more recent restatements. It is worth noting that Gadamer himself noted the parallels between 're-enactment' and his own notion of 'fusion of horizons' (*Horizontsverschmelzung*), which he also sees as resulting from interpretation as opposed to being some alternative 'method' (Gadamer 1999: 374). The title of his book, *Truth and Method* is potentially misleading given his claim that hermeneutics is "not […] a methodology for the human sciences" (Gadamer 1999: xxiii). This raises a similar issue concerning Collingwood: did he have in mind a 'methodology for the human sciences'? This question cannot be properly discussed here, but we should minimally point out that Collingwood distinguished between 'empirical' and 'normative' sciences, which he called 'criteriological', arguing that, at the level of psychology, one must distinguish between, respectively, a 'science of feeling' and a 'science of thought' (EM: 109–11). One can take this to mean that Collingwood had in mind some sort of 'methodological dualism', since 'feelings' at the bottom level could be investigated by the methods of natural sciences (psychophysics), while the domain of 'thought' that includes history and philosophy would require for itself an alternative method. Furthermore, Collingwood's talk of 'Baconian history' as 'scientific' brings him closer to Dilthey and others in that tradition who favoured 'methodological dualism'. In what follows, however, we argue that those who saw him as proposing a 'method of empathetic understanding' or as arguing for a 'rational explanation model' misread him and we refrain from

elevating the parallels with abductive reasoning drawn in Sect. 2 to the level of an alternative 'method'.

4. The Gabbay-Woods Schema is first presented in these, but we will rely here on a more recent presentation in (Woods 2013).
5. For a discussion of this link, see (Marion 2018).
6. There seems to be a confusion in Hempel's text between Dilthey's '*Verstehen*' and Lipps' psychological notion '*Einfühlung*', which indicates that Hempel's knowledge of that tradition was probably superficial and/or second-hand. Hempel does not provide direct sources but refers to *Mandelbaum 1938*, which discusses that very tradition (Dilthey is especially relevant here), albeit in relation to historical relativism, as opposed to the nature of explanation in history. Hempel's other source is Zilsel, who commits the same confusion when writing ""Understanding" means psychological empathy: psychologically a historical process is "understood" if it is evident or plausible" (Zilsel 1941: 577). Hempel also discusses 'understanding' in terms of plausibility (Hempel 1942: 45). This would tend to confirm that Zilsel was his source.
7. The origin of this view is (Donagan and Donagan 1965: 20). The claim is widespread, for example in (Salmon 1992: 232).
8. For a similar claim see (A: 113).
9. The key passage in Collingwood is (IH: 174–75). For a discussion of this point, see (Kobayashi and Marion 2011: 87–92).
10. Among the passages here, see (IH: 215), and see Dray's discussion in (Dray 1995: 52–57), although he does not contrast Collingwood's view with Dilthey's on this very point.
11. It is also called the 'acquaintance theory of knowledge' in (Dray 1957b).
12. Even more so in Donagan's reading, in (Donagan 1966: 150–51). On the 'logical connection argument', see (Stoutland 1970).
13. Reprinted in its original context as (PH: 7–38).
14. It is fitting to note here that to set up imaginary examples and then use our intuitions to reflect on them is a very usual procedure within analytic philosophy.
15. See (A: 30) for the claim that it is archaeology that made him see "the importance of the 'questioning activity' in knowledge". In chapter xi on 'Roman Britain' he extracts three principles from his practice as archaeologist, the first one is his own 'logic of questions and answers', see (A: 122 and 126), and the second one is 're-enactment', although he does not use that word. See (A: 127–28). (The third one concerns 'second-order history' or the 'history of historical thought' about a given problem. It is less important in the context of this chapter.)
16. Within the Hempel-Dray debate broadly construed, Maurice Lagueux is to our knowledge the only one to have argued that the point of an explanation is to remove an initial 'puzzlement'. See (Lagueux 2010: ch. 9). We suggest that Peirce's abduction is a natural extension of this idea.

17. The expression is from (Couse 1990), and parallels have been suggested in (Levine 2004) between Collingwood's *Who killed John Doe?* and Agatha Christie's *The Murder in the Vicarage*. Such parallels are not that important, but for a discussion on the 'detective model', see (Hintikka and Hintikka 1982) and (Marion 2018).
18. See (Aliseda 2017) for a brief overview.
19. There is no room to discuss the appropriateness of the Gabbay-Woods Schema as reading of Peirce, but at least see for example (Campos 2011) or (Mcauliffe 2015) for criticisms of a major rival reading of abduction as 'inference to the best explanation' originating in (Harman 1965). We are also mindful of searching criticisms in (Kapitan 1992) of Peirce's attempt at placing abduction alongside induction and deduction as a *third* form for reasoning.
20. Following here the presentation in (Woods 2013: 369).
21. We skip here a further premise, stating that additional conditions should be added for the acceptability of *H*.
22. The terminology is from (Woods 2013: 371).
23. There is an analysis of this passage in (Hintikka and Bachman 1991: ch. 1), and the case is also discussed in (Marion 2018: 260–68).
24. Collingwood discusses selective excavation in (Collingwood 1931: 36–39). Strangely enough, he does not attribute the method to Simpson and Gibson in his *Autobiography*, but he proudly claims that have succeeded into getting the 'principle of question and answer' adopted at the 1930 Congress of British Archaeological Societies. It is true that to this day no permit would be delivered without any prior explanation of the purpose of the excavation in terms of the questions that hoped to be answered by the excavation.
25. This point is discussed in more details in (Marion 2018: 252–260).
26. This chapter originates in 2010, in a discussion we had with Jan van der Dussen during a meeting of the R. G. Collingwood Society at the Palazzo Datini in Prato. Of course, he is not responsible for its content, but we would like to dedicate this chapter to him, in recognition not only of the help we got from discussing these issues with him, but also of his lifelong achievement in Collingwood scholarship. In philosophy, we always stand on shoulders of giants.

Bibliography

Aliseda, A. 2017. The Logic of Abduction: An Introduction. In *Springer Handbook of Model-Based Science*, ed. L. Magnani and T. Bertolotti, 219–230. Berlin: Springer.

Campos, D.G. 2011. On the Distinction Between Peirce's Abduction and Lipton's Inference to the Best Explanation. *Synthese* 180: 419–442.

Collingwood, R.G. 1931. Hadrian's Wall: 1921–1930. *The Journal of Roman Studies* 21: 36–64.

Conan Doyle, A. 2008. *Complete Works.* Vol. 18. Newcastle-upon-Tyne: Cambridge Scholars.

Couse, G.S. 1990. Collingwood's Detective Image of the Historian and the Study of Hadrian's Wall. *History and Theory* 29 (4): 57–77.

D'Oro, G. 2002. *Collingwood and the Metaphysics of Experience.* London: Routledge.

Donagan, A. 1956. The Verification of Historical Theses. *The Philosophical Quarterly* 6: 193–208.

———. 1966. The Popper-Hempel Theory Reconsidered. In *Philosophical Analysis and History*, ed. W.H. Dray, 127–159. New York/London: Harper and Row.

Donagan, A., and B. Donagan, eds. 1965. *Philosophy of History.* New York: Macmillan.

Dray, W.H. 1957a. *Laws and Explanation in History.* Oxford: Clarendon Press.

———. 1957b. R. G. Collingwood and the Acquaintance Theory of Knowledge. *Revue Internationale de Philosophie* 11: 420–432.

———., ed. 1966. *Philosophical Analysis and History.* New York: Harper & Row.

———. 1995. *History as Re-Enactment.* Oxford: Clarendon Press.

Føllesdal, D. 1979. Hermeneutics and the Hypothetico-Deductive Method. *Dialectica* 33: 319–336.

Gabbay, D.M., and J. Woods. 2005. *The Reach of Abduction. Insight and Trial.* Amsterdam: Elsevier.

———. 2006. Advice on Abductive Logic. *Logic Journal of the IGPL* 14 (2): 189–219.

Gadamer, H.-G. 1999. *Truth and Method.* New York: Continuum.

Gardiner, P. 1952. *The Nature of Historical Explanation.* Oxford: Oxford University Press.

Harman, G.H. 1965. The Inference to the Best Explanation. *Philosophical Review* 74: 88–95.

Hempel, C.G. 1942. The Function of General Laws in History. *Journal of Philosophy* 39: 35–48.

———. 1966. Explanation in Science and in History. In *Philosophical Analysis and History*, ed. W. Dray, 95–126. New York: Harper and Row.

Hempel, C.G., and P. Oppenheim. 1948. Studies in the Logic of Explanation. *Philosophy of Science* 15: 135–175.

Hintikka, J., and J. Bachman. 1991. *What If...? Towards Excellence in Reasoning.* Mountain View: Mayfield.

Hintikka, M.B., and J. Hintikka. 1982. Sherlock Holmes Confronts Modern Logic: Towards a Theory of Information-Seeking Through Questioning. In *Argumentation. Approaches to Theory Formation*, ed. E.M. Barth and J.L. Martens, 55–76. Amsterdam: John Benjamins B. V.

Kapitan, T. 1992. Peirce and the Autonomy of Abductive Reasoning. *Erkenntnis* 37: 1–26.
Kobayashi, C., and M. Marion. 2011. Gadamer and Collingwood on Temporal Distance and Understanding. *History and Theory* 50 (December Theme Issue): 81–103.
Kripke, S. 2017. History and Idealism. The Theory of R. G. Collingwood. *Collingwood and British Idealism Studies. Incorporating Bradley Studies* 23 (1): 9–29.
Lagueux, M. 2010. *Rationality and Explanation in Economics.* London: Routledge.
Levine, J. 2004. The Autonomy of History: R. G. Collingwood and Agatha Christie. In *Re-Enacting the Past. Essays on the Evolution of Modern English Historiography*, 253–264. Aldershot: Ashgate.
Marion, M. 2018. Le modèle interrogatif de l'enquête de Hintikka, la "logique des questions et réponses" de Collingwood et le raisonnement par abduction. *Klesis* 39: 245–269.
Mcauliffe, W.H.B. 2015. How Did Abduction Get Confused with Inference to the Best Explanation? *Transactions of the Charles S. Peirce Society* 51 (3): 300–319.
Peirce Edition Project, ed. 1998. *The Essential Peirce. Selected Philosophical Writings. Vol. 2 (1893–1913).* Bloomington: Indiana University Press.
Ricœur, P. 1984. *The Reality of the Historical Past.* Milwaukee: Marquette University Press.
Salmon, M.H. 1992. Philosophical Models for Post-Processual Archaeology. In *Metaarchaeology. Reflections by Archaeologists and Philosophers*, ed. L. Embree, 227–241. Dordrecht: Kluwer.
Scriven, M. 1959. Truisms as the Grounds for Historical Explanations. In *Theories of History*, ed. P. Gardiner, 443–475. New York/London: Macmillan.
Stoutland, F. 1970. The Logical Connection Argument. In *Studies in the Theory of Knowledge, American Philosophical Quarterly, Monograph Series n. 4*, ed. N. Malcolm et al., 117–129. Oxford: Basil Blackwell.
van der Dussen, J. 1981. *History as a Science. The Philosophy of R. G. Collingwood.* The Hague: Martinus Nijhoff.
———. 2007. Collingwood's Claim that History Is a Science. In *Collingwood and British Idealism Studies*, vol. 13/2, 5–30; Reprinted in J. van der Dussen, *Studies on Collingwood, History and Civilisation*, 137–152. Cham: Springer.
Walsh, W.H. 1967. *An Introduction to Philosophy of History.* London: Hutchinson.
Woods, J. 2013. *Errors of Reasoning. Naturalizing the Logic of Inference.* London: College Publications.
Zilsel, E. 1941. Physics and the Problem of Historico-Sociological Laws. *Philosophy of Science* 8: 567–579.

CHAPTER 11

Collingwood and Archaeological Theory

Stephen Leach

It is clear from his obituaries that Collingwood was, to many of his contemporaries, better known for his work in history and archaeology than for his work in philosophy.[1] Indeed, he was widely acknowledged as the foremost authority on Roman Britain. He referred to his work in history and archaeology as the work of his 'laboratory' (A: 121), as opposed to his work in philosophy, which he referred to as the work of his 'study' (ibid). He believed that, as a philosopher who wrote about other academic disciplines, he should have first-hand knowledge of the work carried out within those disciplines.

It might well be argued that this message has still to be heard; but perhaps it was inevitable that Collingwood's stand against the trend of increasingly narrow specialisation in academia would have limited success.

Yet, it is surely of interest to those interested in Collingwood's philosophy that over more than the last 40 years a sub-discipline has flourished within archaeology, archaeological theory,[2] which has itself taken an interest in Collingwood's work. At the foundation of this sub-discipline is Collingwood's first principle of historical thinking, the "'logic of question and answer' as applied to archaeology" (A: 122): the data yields its answers

S. Leach (✉)
School of Politics, IR and Philosophy, Keele University, Staffordshire, UK
e-mail: s.d.leach@keele.ac.uk

K. Dharamsi et al. (eds.), *Collingwood on Philosophical Methodology*, Philosophers in Depth,
https://doi.org/10.1007/978-3-030-02432-1_11

in relation to the acuity of the investigator's questions; the investigator must therefore be aware of the questions that are posed in the course of an investigation, and should at no time dig 'blind' in the hope that the evidence will simply tell its own story. Collingwood was satisfied that this principle had taken root in practice, but its roots are by now even more deep-rooted, and, as will be seen, it might be argued that archaeological theory has flowered as a result.

So—with reference to Collingwood's metaphilosophy—it is worth asking: how would Collingwood have viewed archaeological theory? What would he have thought of the development of this study-cum-laboratory—theory that has developed entirely within archaeology and which is orientated solely towards the solution of archaeological problems? And, indeed, what is the relationship between Collingwood's metaphysics and archaeological theory?

I argue in this chapter that Collingwood would have welcomed the development of archaeological theory but nonetheless, he is himself primarily a metaphysician rather than an archaeological theorist.

1 The Oversights of 'Minute' Philosophers

To have a deep first-hand knowledge of more than one academic discipline is today very rare and was already rare in Collingwood's day. In this respect, in having an archaeological and historical 'laboratory', Collingwood was unusual. It was due to increased specialisation that a 'revolution in historical method' (A: 81) occurred without being noticed by philosophers. The revolution consisted in the development of the method, or the 'logic', of question and answer. 'Scissors and paste' history—in which the historian pieces together the evidence like the leaves of a fallen manuscript—was rejected in favour of the development of the historian's self-consciousness with regard to the questions that might be asked of the evidence. The logic of question and answer is such that, following the revolution, the direction of the inquiry is dictated by the interaction of the historian's questions and the answers supplied in response to them, by a reading of the evidence.

> The historian has to decide exactly what it is that he wants to know; and if there is no authority to tell him, as in fact (one learns in time) there never is, he has to find a piece of land or something that has got the answer hidden in it, and get the answer out by fair means or foul. (ibid.)

The 'revolution in historical method' was led by, among others, Collingwood's mentors Francis Haverfield and David George Hogarth.

> Haverfield himself, least philosophical of historians, cared nothing about the principles or the potentialities of the revolution he was leading. He never even seemed aware that a revolution was going on. (A: 83)

Likewise, that 'bold revolutionary' (A: 82) David George Hogarth had no interest in philosophy.[3] Collingwood's point is not that Haverfield and Hogarth should have taken an interest in philosophy but rather that philosophers should have taken more notice of this 'Baconian revolution' (A: 115) in the study of history. It should not have entirely escaped their notice, especially in Oxford where undergraduates studying 'Greats' were required to study both philosophy and history.

Collingwood did not argue that all philosophers should have the luxury of their own laboratories as he did, but he strongly suggested that if they must seclude themselves in their studies they should at least now and then look out of the window. Collingwood's metaphysics involves: first, becoming familiar with what is, by common consent, good practice in other disciplines that take place outside of the philosophical study; and then, second, but crucially, conducting a purely metaphysical ('armchair') inquiry into the necessary logical presuppositions of those practices.

With increasing academic specialisation ever since Collingwood's day, it is not surprising that his example has not been widely imitated. However, if we have any sympathy at all for Collingwood's position, then, lest we repeat the mistake of the narrowly focused 'minute' philosophers of Collingwood's youth (A: 15–21),[4] it is worthy of some notice that over more than the last 50 years or so, a sub-discipline has been born and flourished within archaeology—archaeological theory. The development of this sub-discipline has been almost entirely ignored by those philosophers who are interested in Collingwood's work. This oversight is, I believe, comparable to the oversight of which Collingwood complained, when the philosophers of his student years ignored developments in the discipline of history. It is all the more remarkable in that, conversely, many archaeological theorists have read some of Collingwood's work.

2 Collingwood in Archaeological Theory

In browsing books on archaeological theory, it will not be long before the reader comes across references to Collingwood. The reader will not find any deep exposition of Collingwood's theories in these books but it will soon become clear that in archaeological theory Collingwood is seen as more than just an illustrious ancestor. He is someone about whom opinion is divided.

At this point, in order to understand the geography of archaeological theory, it is necessary to know a little of its history. In broad terms, archaeological theory refers to the theory through which archaeologists study their data—in other words, archaeological theory is used 'to get the answer out'. The term first came into widespread use in the 1960s with the rise of 'New Archaeology' (or processual archaeology). Of course, it was not the case that before then all archaeologists believed in the power of the innocent naked eye; but in the 1960s and 1970s there was a new confidence in what an awareness of one's own viewpoint might achieve.

New archaeologists distanced themselves from what has become known as culture history—the view that cultures are invariably autonomous, discrete and autonomous units each with their own unique character, interacting much like individuals. New archaeologists, such as Lewis Binford, David Clarke, and Colin Renfrew believed that archaeology should be closely allied with physical anthropology. At the extreme end of New Archaeology's spectrum, it was believed that the only difference between archaeology and physical anthropology is that physical anthropologists study present-day artefacts whereas archaeologists study ancient artefacts; however, it was believed that that difference was not, at a fundamental level, theoretically significant. In broad terms, New archaeologists believed they could learn far more about underlying historical processes if they modelled archaeology on the natural sciences.

The next wave of archaeological theory was the unimaginatively named post-processual archaeology. It was led by archaeologists such as Michael Shanks, Christopher Tilley, and Ian Hodder. This movement arose in the late 1970s and 1980s but its heyday in terms of influence was arguably in the 1990s. However, there was considerable chronological overlap between the two movements. The post-processualists stressed the importance of historical context, which they believed New Archaeology ignored. Two contexts may share certain features but they are never exactly alike, nor can we assume similarities between a past society and our own. Thus, an artefact that at first glance seems to be similar to something in our own society may on closer inspection, and with a greater appreciation of historical context, be revealed to have a radically different purpose. Post-processualists argued that investigation involves acclimatising ourselves to the unfamiliar and appreciating alien rationales. Sensitivity to context is paramount. We should not assume that we exist in any form of deterministic relationship with our environment. Our actions are influenced by our thoughts about our environment. (Yet that is not to say that the post-processualists were in favour of culture history. They were just as opposed to culture history as the processualists.)

Since the 1980s and the 1990s the vigour and the volume of debates have slowly diminished. There has even recently been talk of the 'death of theory' (see Bintliff and Pearce 2011) whilst others talk, more optimistically, of a new pragmatism in archaeological theory (see Preucel and Mrozowski 2010) in which it is unnecessary to be irrevocably committed to just a single approach. Yet it is still the case that archaeological theorists tend to divide between those who would ally archaeology with the social sciences and those who would ally archaeology with the natural sciences.

Within these debates, although Collingwood's work has not been discussed in depth he has been cited with approval by post-processualists and their sympathisers and with opposing disapproval by processualists and their sympathisers. In this regard, the following exchange between the post-processualist Ian Hodder and John Bintliff is typical. Hodder stressed the importance of reconstructing historical agents' motivations in order to understand their actions, and he takes Collingwood to have advocated a particular 'technique' for doing this (Hodder 1991: 98). Bintliff, however, was critical of his attempts "to resurrect the increasingly popular 'empathy' techniques of the pre-war historian and philosopher R.G. Collingwood. In this approach […] one immerses oneself in all the contextual material for a particular event or circumstance – then, using one's imagination and ability to step into the mental shoes of the relevant historical actors, one is able to make explicit why certain decisions were made, why certain things happened, through the medium of reconstructing the thought processes of past individuals" (Bintliff 1988: 13). Bintliff criticised the underlying assumption of this technique that "past personages are like you and me" (Bintliff 1988: 14). (Note that here Bintliff is turning a criticism usually made of New Archaeology against the post-processualist Ian Hodder.)

Collingwood scholars will by this point have recognised an all too familiar error in the interpretation of Collingwood's work. Both Hodder and Bintliff repeat the error originally made by, among others, the philosopher Patrick Gardiner in his criticism of Collingwood. According to Gardiner:

> When in doubt as to why a person did something, 'put yourself in his position' may be a useful methodological precept, but it is far from always being reliable. People differ. From the fact that, if I did *x*, it would be because I wanted *y*, i.e. would have been satisfied if *y* occurred, would have given *y* as my aim if asked, &c., it does not follow that when a medieval baron did *x* he wanted *y*. (Gardiner 1952: 129–30)[5]

This misunderstanding of Collingwood has been put about ever since the publication of his most widely read work *The Idea of History* in 1946. It was a mistake encouraged by the fact that *The Idea of History* was compiled posthumously in a misleading form, a form that its author never intended.

Collingwood's primary intention was not prescriptive. His intention, as Collingwood scholars have explained repeatedly, was to isolate the distinctive logical structure of what by common consent are recognised as successful historical explanations. Thus, Collingwood would agree with Gardiner and Bintliff: we cannot assume that "past personages are like you and me". His main point is subtly but crucially different. It is a conceptual rather than a prescriptive point.[6] It is that insofar as the historian claims to have understood an agent's actions the historian will claim to have understood, albeit at a rudimentary level, the premise of the agent's reasoning.

However, that Collingwood's *primary* position is not prescriptive was not pointed out to Hodder and Bintliff—and among archaeological theorists, this misunderstanding of Collingwood's position persists.

There is—incidentally—a second error in Hodder and Bintliff's interpretation of Collingwood. They assume that Collingwood's conception of *archaeology* (of its distinguishing characteristics)—as opposed to history—is the same as theirs. It is not.

In the preface to his *The Archaeology of Roman Britain* (1930), a work that classifies all of the various Roman finds and Roman sites that might be encountered within Britain, Collingwood is emphatic that this "is strictly [...] a handbook of archaeology, not a history" (ARB: vi). Alternatively, in Collingwood's vocabulary, *The Archaeology of Roman Britain* might be described as a work of 'critical history', for, as he makes clear in the following passage, in his view 'critical history' is synonymous with archaeology.

> Critical history classifies its sources into groups, and then subdivides these groups, and then subdivides these groups, framing rules for the manipulation of the various subdivisions. Taken as a whole, this technique is an abstract or

> classificatory science, which has no general name, unless that of archaeology is used for it, and is subdivided into numerous departmental sciences such as palaeography, numismatics, epigraphy and so forth. They are not themselves history; they are only methods of dealing with the sources of history. (IH: 490)[7]

The idea is that history in a rudimentary form is present in archaeological reports, but these summations are then handed to historians for further processing. Archaeology, in this respect, is 'historical' in that it serves the historian.

Admittedly Collingwood does not discuss the distinction between archaeology and history at any great length. Perhaps he planned to address the question in his unfinished *Principles of History* (1999) but he was more interested in the Baconian revolution in *history* than any other. He believed that it was *this* revolution that was driving developments in archaeology. On the evidence of the preface to *The Archaeology of Roman Britain*, he believed that archaeology—as distinct from history—was more orientated towards such classificatory questions as 'what is this?', 'what happened?', and 'when?' whereas the question of 'why?' distinguished the historian's domain. Collingwood believed that it was the historian's question of 'why?' and the distinctive manner in which this question was answered which differentiated the discipline of history from the domain of natural science. However, it is debatable as to whether he would make the same claim for archaeology as he would for history.

Hodder, by contrast, would have it that the archaeologist, as much as the historian, is focussed on the question 'why?' and on the reconstruction of an agent's reasoning. He takes Collingwood to be his ally; however, any reader would be hard-pressed to find an explanation of 'why?' in Collingwood's 'handbook of archaeology' *The Archaeology of Roman Britain* of the sort that reconstructs the motivating premise of an agent's reasoning. Reconstruction of the agent's premise is rather something that is found in Collingwood's *history* of Roman Britain when, for example, he discusses Caesar's invasion plans.

As mentioned, the differences between history and archaeology were not fully explored in Collingwood's work, nonetheless they are there in his work. This is an example of an area that might usefully be further explored with the help of Collingwoodian metaphysics.[8]

But to return to the main point (of this section): archaeological theorists have not entirely ignored Collingwood but they have tended to see him in their own image (in the case of post-processualists and their successors) or in the image of their opponents (in the case of processualists and their successors). In doing so, they have repeated mistakes in interpreting Collingwood's philosophy of the sort that were originally made, and corrected,[9] in the 1950s and 1960s.

3 Three Principles

It has been argued that in his philosophy of history Collingwood's primary intention was to isolate the distinctive logical structure of what by common consent are recognised as successful historical explanations. In accordance with the theory of presuppositions outlined in *An Essay in Metaphysics* (1940), his ultimate objective was to isolate the absolute presuppositions of the discipline of history. That is to say, he wished to isolate those presuppositions that are logically essential to the discipline, and which give the discipline its essential characteristics, though yet they may very well be unquestioned by historians themselves (Note incidentally that: "The logical efficacy of a supposition does not depend upon the truth of what is supposed, or even on its being thought true, but only on its being supposed") (EM: 28).

This exercise he never completed. It was to have been achieved in *The Principles of History* (1999) but this work was incomplete at the time of Collingwood's death (and that part of it which existed in the manuscript was for many years lost). Nonetheless, his investigations had progressed far enough that he was able to make the case that historians' explanations have a different and distinctive logical structure than the explanations of natural scientists. Historians claim to have understood an action when they are able to reconstruct the agent's motivating premise.

Yet, having emphasised that Collingwood's *primary* intention was not to provide a prescriptive methodology, it must be admitted that there is a recessive prescriptive aspect to Collingwood's work. This is owing to the fact that in accordance with his theory of presuppositions, he is in effect holding up a mirror towards the best practices in "historical thinking" (A: 121).[10] (There is an unavoidable element of circularity to this process in that at the start of the investigation it is the metaphysician holding the mirror who chooses best examples of good practice. Collingwood admits this: "The metaphysician must himself make presuppositions" (EM: 63). However, this is a virtuous circularity in that it enables inquiry to get off the ground and in that the prudent metaphysician will rely on a broad consensus of what constitutes best practice.) It is in holding up a mirror to best practice—and making explicit that which was previously taken for granted—that there arises the prescriptive aspect in Collingwood's metaphysics.

This aspect is most notable in his *Autobiography* in which he advocates the following three principles of "historical thinking" (A: 121).

1. "Long practice in excavation had taught me that one condition – indeed the most important condition – of success was that the person responsible for any piece of digging, however small and however large, should know exactly why he was doing it. He must first of all decide what he wants to find out, and then decide what kind of digging will show it to him. This was the central principle of my 'logic of question and answer' as applied to archaeology." (A: 122)
2. "A second principle was that, since history proper is the history of thought, there are no mere 'events' in history: what is miscalled an 'event' is really an action, and expresses some thought (intention, purpose) of its agent; the historian's business is therefore to identify this thought." (A: 127–28)
3. "A third principle was that no historical problem should be studied without studying what I called its second-order history; that is, the history of historical thought about it." (A: 132)

It is noticeable that in these principles history and archaeology are not distinguished. As we have seen, Collingwood paid very little attention to their differences. It is clear that he assumed that history and archaeology are distinct but he did not focus on their differences in his metaphysical analyses. Archaeology features in the first principle of historical thinking because it serves history and because archaeologists frequently are, in fact, motivated by questions taken from the discipline of history. Collingwood's focus was rather on the distinction between history and the natural sciences.

Taking them in reverse order, *the third* of these principles is not in doubt, and, furthermore, with the advent of the internet, is now easier to respect in practice than ever before. Site reports can often be read at home in front of the computer. A special trip to the archives may no longer be required. The third principle is in fact so well established that it is worth reminding ourselves of why Collingwood thought it worthwhile to make this point. He is contrasting modern techniques to those which have been superseded. In the hope of consolidating the progress which he believes has been made in history, he is aiming to make those principles on which effective investigation relies fully explicit. In brief, it is because the data of the investigation yields answers in relation to the acuity of the investigator's questions that it is worth studying the history of historical thought about a problem. One may thereby become more aware of one's own inherited presuppositions and one may also come to see the data in a new light and so notice the significance of a pattern that had previously been overlooked.

Collingwood might have at this point gone on to discuss the important distinction, drawn by both archaeologists and historians between data and interpretation. This is a distinction that is not found in the work of 'scissors-and-paste' historians but which is found in the work of all contemporary academic historians and archaeologists. It is a distinction that is made in recognition of the fact that future generations of archaeologists and historians may wish to revisit old data-sets and perhaps to reinterpret them in the light of new questions and with the aid of new techniques. Thus, General Pitt-Rivers, of the generation prior to Francis Haverfield, has been praised by later archaeologists such as Mortimer Wheeler and Christopher Hawkes (Collingwood's pupil), for keeping records of his excavations of such a standard that they could be reinterpreted by later generations. In short, he is admired in large part because he left accurate records even of that which he did not understand. But perhaps here we are at a level at which there is a common ground between the humanities and the natural sciences. For in the natural sciences too, the distinction drawn between data and interpretation is fundamental to good practice.

The second principle was the focus of a great deal of Collingwood's attention throughout his work in the philosophy of history. The claim that "all history is the history of thought" is frequently repeated throughout his work (A: 110; IH 215, 317; PH 67, 98. 100). In his *Autobiography* he cites as an example of the second principle, the question of 'what was Hadrian's Wall for?' Surprisingly, Collingwood was the first to ask this question. Previously, the Wall had simply been categorised, vaguely, as "a frontier defence" (A: 128), but no one had asked how it worked. Collingwood argued that it was not efficient as a purely defensive monument, from which attacks might be repelled. It might better be imagined as an elevated 'sentry-walk' with panoramic views over the approaches from the north. It was by this means that the Roman army held dominion over the surrounding land. (In recent years, following the work of Paul Bidwell (2005), there has been a revival of the idea of the Wall as a purely defensive barrier but this conception has supplemented rather than displaced Collingwood's conception of the Wall.) Collingwood's point was that historians must ask such questions as will enable them to recover the relevant premise behind an action, that from which historical agents reasoned.

Incidentally, Collingwood also points out how his new explanation of Hadrian's Wall led him to revisit the data of earlier archaeologists:

> A question answered causes another question to arise. If the Wall was a sentry-walk, elevated from the ground and provided (no doubt) with a parapet to protect the sentries from sniping, the same sentry-walk must have continued down the Cumberland coast, beyond Bowness-on-Solway, in order to keep watch on vessels moving in the estuary; for it would have been very easy for raiders to sail across and land at any unguarded point between Bowness and St. Bee's Head. But here the sentry-walk need not be elevated, for sniping was not to be feared. There ought, therefore, to be a chain of towers, not connected by a wall but otherwise resembling those on the Wall, stretching down that coast. The question was, did such towers exist?
>
> Search in old archaeological publications showed that towers of exactly the right kind had been found; but their existence had been forgotten, as generally happens with things whose purpose is not understood. Search on the ground in 1928 revealed a number of other places where it seemed possible that others might yet be revealed by future excavation. (A: 129–30)

Yet, though Collingwood makes the point that "all history is the history of thought" would he also make the point that "all archaeology is the archaeology of thought"? Perhaps only for the reason that archaeology serves history? As already noted, it is regrettable that he did not discuss this distinction at any length.[11]

Collingwood's main focus is on the distinction between history and the natural sciences, and his three principles are specifically principles of "historical thinking" (A: 121). However, it is interesting that it is in archaeology—and, in particular, archaeological theory—that Collingwood's *the first principle* is most frequently and explicitly acknowledged.

Collingwood did not claim originality for the first of his principles—though it was only Collingwood who controversially described it as a 'logic' (A: 29–43). He credited his teachers Haverfield and Hogarth in regard to the first principle: but it is already to be found in Flinders Petrie's *Methods and Aims in Archaeology* (1904):

> The old saying that a man finds what he is looking for in a subject is too true; or, if he has not enough insight to ensure finding what he looks for, it is at least sadly true that he does not find anything that he does not look for. (Flinders Petrie 1904: 49)

In other words, it is a necessary, but not sufficient, condition in archaeological investigation, to have some question in mind to which one seeks an answer.

Of his contemporaries, Collingwood mentions R.E.M. Wheeler as a fellow advocate of this principle and he credits Wheeler for embedding it within the curriculum of the London Institute of Archaeology (A: 126). Collingwood, looking back to the 1920s in his *Autobiography* (1939), notes how well the principle has taken root. Consequently: "About the future of this principle among scholars I am [...] not anxious" (A: 126–27).

His confidence was justified. A seemingly ineradicable distrust of theory (of any kind) still prevails in some parts of archaeology (as it also does in history), but the principle has nonetheless grown ever more deeply rooted. The attitude that 'the evidence speaks for itself', without any question asked of it, was the attitude that Collingwood frequently encountered at the Society of Antiquaries in the 1920s, but to encounter that attitude today among those with responsible positions would be unusual.

Indeed, archaeological theory can be seen as brought about by the health of this principle, for to conceive of archaeological theory without it is, I believe, not possible. It is theory that encourages an awareness of our questions. It may also be fundamental to the discipline of history, but the discipline of history has not created an equivalent to archaeological theory. Since Collingwood's day, archaeologists have tended, in general, to be more welcoming of theory than have historians.

4 Would Collingwood Have Approved of Archaeological Theory?

I believe that Collingwood would have approved of archaeological theory for, after all, it has grown from the first of his three principles; but that does not imply that we should see Collingwood as an archaeological theorist.

Although there is a prescriptive aspect to his work, Collingwood asks different questions and has different interests from the archaeological theorist. As, first and foremost, a metaphysician without ontology—that is to say as a metaphysician investigating the structures of knowledge rather than the nature of being—he asks himself about the absolute (fundamental) presuppositions of a discipline. As a metaphysician, Collingwood keeps an objective distance from his work as an archaeologist and historian. He observes this work with great interest, but there is no merging of either of these disciplines with metaphysics or vice versa.[12] When Collingwood investigates the foundations of history or archaeology it is not *primarily* for the sake of advancing those disciplines. He investigates them because, like Kant, he is interested—and perhaps for the sake of interest alone—in the logical foundations of knowledge. Indeed, insofar as there is also, unavoidably, a prescriptive aspect to this work it will be more convincing for it not being Collingwood's primary intention.

By contrast, the archaeological theorist *is* primarily empirically driven—asking, first and foremost: what will best advance inquiry? ("Archaeological theory is not independent of the problems that need to be solved: it arises out of particular problems and articulates them with others" (Yoffee and Sherratt 1993: 1)).

Herein lies the main difference between Collingwood's metaphysics and archaeological theory.

5 Concluding Remarks

A student recently remarked to me that although Collingwood's work was impressive and interesting his followers seem to be taking a long time sharpening their pencils before actually getting down to the business of following their master's example and actually trying to recover the absolute presuppositions of different disciplines.[13] I had to agree. Perhaps one reason is that philosophers who advocate Collingwood's metaphysics without ontology remain in a minority position. In this position, it is perhaps unsurprising that Collingwood's supporters concentrate on promoting his conception of metaphysics rather than studying other disciplines as he recommended.

However, although there *is* still work to be done explicating Collingwood's metaphilosophical position, that should not divert us from the task—left unfinished by Collingwood—of recovering the distinctive presuppositions of different disciplines. Collingwood made an excellent start on this work, for example in his isolation of different senses of 'cause' distinctive of natural science and of history (EM: 285–327), but, as I have tried to show, there are still areas in which there is work to be done. In short, we need to sometimes look out of the study window and to pay attention to developments in other disciplines. One of the most noteworthy is the development of archaeological theory. Although he cannot himself be classed as an archaeological theorist, Collingwood would surely have found this development to be both gratifying and of great interest.

In the other direction, it may also be of interest, and provide some reassurance and backing to archaeological theorists—in defending their subdiscipline from colleagues who may be more mistrustful of theory—to learn that from the metaphysician's perspective both archaeological fieldwork and archaeological theory are fundamentally part of practice. For archaeological theory, is *not* primarily theoretical—in the manner of Collingwood's metaphysics without ontology. Archaeological theory is empirically driven, whereas Collingwood's metaphysics is not.[14]

Notes

1. See, for example, *The New York Times*, 12 January 1943, 'Dr. Collingwood, Oxford Professor: Noted Authority on Life in Britain under Romans Dies at Age of 53'.
2. Sometimes referred to as theoretical archaeology.
3. According to Caroline Barron (Hogarth's grand-daughter) (personal comment). There is something mischievous about Collingwood's description of Haverfield and Hogarth as revolutionaries.
4. Minute in the sense of small. The term was used by Cicero to describe philosophers who denied the immortality of the soul. It was taken up in Berkeley's *Alciphron* (or the Minute Philosopher) and it also occurs in Reynolds' *Discourses.*
5. See also (Popper 1972: 173; 1963: 38, n. 4).
6. This is overlooked in Hodder (1991, 1995).
7. See also (IH: 258–9) and (PH: 14).
8. In (Leach 2017) I have explicated a distinction between archaeology and history that may be compatible with Collingwood's views. I there defend the views of the Russian archaeological theorist Leo S. Klejn. Klejn argues that archaeology is closely related to detective work (which he argues is neither one of humanities nor a natural science but an autonomous branch of knowledge) and that it is focussed upon such questions of 'what?' 'when?' and 'where?' whereas history is a humanistic discipline focussed upon the question 'why?' Archaeologists feed the results of their inquiries to historians for further processing. However, that is not to say that either discipline is ultimately superior or inferior to the other. As Klejn points out, archaeology might equally well be described as the servant of history or as

its mother. The two disciplines have close 'business relations'. However, although the same person may be both an historian and an archaeologist, and even though the archaeologist may be stimulated by the historian's questions, the two disciplines remain conceptually distinct. I argue that this model is not at odds with Collingwood's philosophy of history. Indeed, Klejn agrees with Collingwood's view of historical explanation.

9. Corrected in, e.g., (Dray 1958, 1963).
10. From the prescriptive aspect to Collingwood's work it follows that if Collingwood were to have been a consistently bad field archaeologist this might cast doubt on his work as a theorist. Largely on the basis of Collingwood's excavations of the Cumbrian henge monument, King Arthur's Round Table, Richmond (1943) and Hodder (1995) suggest that Collingwood was weak in this area. However, this verdict is by no means generally accepted. See Simpson (1998).
11. Another more recent field that awaits exploration by the metaphysician without ontology is the discipline of material culture studies. It has developed as an academic discipline since the 1990s. According to the *Journal of Material Culture*: "It is concerned with the relationship between artefacts and social relations irrespective of time and place and aims to systematically explore the relationship between the construction of social identities and the production and use of culture." http://journals.sagepub.com/home/mcu [accessed 23.12.17]. If archaeology is most closely related to detective work, then perhaps material culture studies (less focused on the specifics of 'what happened?' 'when?' and 'where?) is related to forensic science?
12. 'For the most part' because there is one conspicuous exception: his chapter on 'Art' in *Roman Britain and the English Settlements.* In his *Autobiography* Collingwood refers to this as "a chapter which I would gladly leave as the sole memorial of my Romano-British studies, and the best example I can give to posterity of how to solve a much-debated problem in history, not by discovering fresh evidence, but by reconsidering questions of principle." (1939: 144–45). Collingwood's other insights as an archaeologist and historian have been incorporated into subsequent work, although sometimes with a degree of modification, or else they are discussed but still held in abeyance. By contrast, his work on the revival of Celtic art has been largely, and in my view rightly, ignored. Whereas Collingwood usually worked as a critical, or analytical philosopher of history, concerned with the logical structure of historian's explanations, in 'Art' he worked as a speculative philosopher, viewing the unfolding of events as itself an *essentially* logical process.
13. My thanks to Aaran Burns.
14. Thanks to Karim Dharamsi and Giuseppina D'Oro for their helpful suggestions on this chapter.

Bibliography

Bidwell, P. 2005. The System of Obstacles on Hadrian's Wall: Their Extent, Date and Purpose. *Arbeia Journal* 8: 53–75.

Bintliff, J., ed. 1988. *Extracting Meaning from the Past*. Oxford: Oxbow.

Bintliff, J., and M. Pearce, eds. 2011. *The Death of Archaeological Theory*. Oxford: Oxbow.

Collingwood, R.G. 1930. *The Archaeology of Roman Britain*. London: Methuen.

———. 1939. *An Autobiography*. Oxford: Oxford University Press.

———. 1940. *An Essay on Metaphysics*. Oxford: Oxford University Press.

———. 1946. *The Idea of History*. Ed. T.M. Knox. Oxford: Oxford University Press.

———. 1999. *The Principles of History*. Oxford: Oxford University Press.

Dray, W.H. 1958. Historical Understanding as Rethinking. *University of Toronto Quarterly* 27: 200–215.

———. 1963. Historical Explanation of Actions Reconsidered. In *Philosophy and History*, ed. S. Hook. New York: New York University Press.

Flinders Petrie, W.M. 1904. *Methods and Aims in Archaeology*. London: Macmillan.

Gardiner, P. 1952. *The Nature of Historical Explanation*. London: Oxford University Press.

Hodder, I. 1991. *Reading the Past: Current Approaches to Interpretation in Archaeology*. 2nd ed. Cambridge: Cambridge University Press.

———. 1995. Of Mice and Men: Collingwood and Archaeological Thought. In *Philosophy, History and Civiliation: Interdisciplinary Perspectives on R.G. Collingwood*, ed. D. Boucher, J. Connelly, and T. Modood. Cardiff: University of Wales Press.

Leach, S. 2017. Leo S. Klejn and R.G. Collingwood on History, Archaeology and Detection. *Journal of the Philosophy of History* 11: 391–407.

Popper, K. 1963. *Conjectures and Refutations: The Growth of Scientific Knowledge*. London: Routledge.

———. 1972. *Objective Knowledge: An Evolutionary Approach*. Rev. ed. Oxford: Oxford University Press.

Preucel, R.W., and S.A. Mrozowski, eds. 2010. *Contemporary Archaeology in Theory: The New Pragmatism*. 2nd ed. Chichester: Wiley-Blackwell.

Richmond, I.A. 1943. Appreciation of R.G. Collingwood as an Archaeologist. *Proceedings of the British Academy* 29: 476–485.

Simpson, G. 1998. Collingwood's Latest Archaeology Misinterpreted by Bersu and Richmond. *Collingwood and British Idealism Studies* 5: 109–119.

Yoffee, N., and A. Sherratt. 1993. *Archaeological Theory: Who Sets the Agenda*. Cambridge: Cambridge University Press.

Index[1]

[1] Note: Page numbers followed by 'n' refer to notes.

K. Dharamsi et al. (eds.), *Collingwood on Philosophical Methodology*, Philosophers in Depth,
https://doi.org/10.1007/978-3-030-02432-1

The manufacturer's authorised representative in the EU is Springer Nature Customer Service Centre GmbH, Europaplatz 3, 69115 Heidelberg, Germany. If you have any concerns regarding our products, please contact ProductSafety@springernature.com

Printed and bound by CPI Group (UK) Ltd, Croydon, CR0 4YY
01/10/2026
02226249-0004